Library of
VOID
Davidson College

Geometric Optics

The Matrix Theory

GEOMETRIC OPTICS

The Matrix Theory

J. WARREN BLAKER
Department of Physics
Vassar College
Poughkeepsie, New York 12603

MARCEL DEKKER, INC. New York   1971

COPYRIGHT © 1971 by MARCEL DEKKER, INC.

ALL RIGHTS RESERVED

No part of this work may be reproduced or utilized in any form or by any means, electronic or mechanical, including *Xeroxing, photocopying, microfilm, and recording*, or by any information storage and retrieval system, without permission in writing from the publisher.

MARCEL DEKKER, INC.
95 Madison Avenue, New York, New York 10016

LIBRARY OF CONGRESS CATALOG CARD NUMBER: 73-157832
ISBN NO.: 0-8247-1046-0

PRINTED IN THE UNITED STATES OF AMERICA

for

Cindy

Preface

For the past five years I have taught the geometric optics part of the undergraduate optics course at Vassar using the matrix approach. This experience has been most satisfying. Students have been encouraged to make use of the matrix theory they have previously learned and have, I think, found this general approach to optical systems more exciting than the usual geometric treatment. One of the most useful attributes of this approach rests in the use of ray transformations which gives strong physical insights into the properties of optical systems in addition to providing experience in matrix transformations so useful in the students' later work. It is my hope that this approach will find widespread use among physicists within the next few years.

The thin lens law is developed traditionally in Chapter 1 as a general review. In Chapter 2 the ray transformations are developed and then applied to the general system in Chapter 3 where imaging is also discussed. Chapter 4 deals with the cardinal points of systems, graphical construction, and contains a demonstration of the consistency of the matrix approach through a second derivation of the thin lens law. The extension to reflecting surfaces is carried out in Chapter 5. Stops and pupils are discussed in Chapter 6 as well as chromatic aberration since both of these topics affect the quality of the image and both can be treated within the paraxial-ray assumptions. In the final chapter higher-order aberrations are discussed and a general consideration of a number of common optical instruments is undertaken. A number of problems are given with each chapter.

Several of my students have been most helpful in the development of this work. They have made many organizational comments which I trust have led to a smoother treatment.

I particularly want to express my thanks to Professor Stanley Ballard for his friendship, encouragement, and helpful discussions over the past decade as well as for his reading of several sections of the manuscript.

I also want to thank Professor B. R. Coles of Imperial College, University of London for his hospitality during the 1969–70 year during which time part of the writing was done.

Poughkeepsie, New York
June 1971

J. WARREN BLAKER

Contents

Preface   v

Chapter 1. **Introduction**   1

Fermat's Principle   2
Reflection and Refraction   4
Refraction at a Spherical Surface   7
The Thin Lens   10
Images Formed by Reflection   11
Problems   13

Chapter 2. **Refraction and Translation Matrices**   15

Sign Conventions   15
The Refraction Matrix   16
Translation Matrix   19
The Paraxial-Ray Assumption   21
Problems   26

Chapter 3. The Optical System; Imaging   27

Imaging   30
Problems   34

Chapter 4. Additional Properties of the System Matrix   37

Unit Points and Unit Planes   37
Focal Points and Focal Planes   39
Nodal Points   43
Graphical Image Construction   45
The Thin Lens   47
Changing the External Medium   49
Problems   51

Chapter 5. Mirrors   53

The Plane Mirror   56
The Convex Mirror—An Example   57
A Lens with a Reflecting Surface   58
Problems   61

Chapter 6. Stops and Pupils; Chromatic Aberration   63

Stops and Pupils   63
Vignetting   69
Chromatic Aberrations   70
Problems   75

Chapter 7. Geometric Aberrations   77

Spherical Aberration   80
Coma   83
Astigmatism   84
Curvature of the Field   87
Distortion   87
Higher-Order Aberrations   88

Chapter 8. Optical Instruments   89

The Simple Magnifier   89
Eyepieces   92
Telescopes   95
The Microscope   99
Cameras and Camera Lenses   101

Projection Systems	105
Problems	105

Appendix 1 **Matrices**   107

Special Matrices	111
A Useful Theorem	112

Appendix 2 **Mathematical Theory of Aberrations**   115

Bibliography   123

Index   125

Geometric Optics

The Matrix Theory

Chapter 1

Introduction

    The study of the nature of light in the fullest detail is an extremely difficult and complex undertaking. Fortunately, there are various simple models available which can be used to examine problems of special interest. These models are quite reliable in their own somewhat limited regions of applicability. The three most significant and the three oldest of these models are the geometric or ray model, the wave model, and the particle or quantum model. In this book we concern ourselves with only the first of these, viz., the geometric or ray model.

    The ray model is based on the fact that light propagating from a point source in a uniform medium produces shadows which appear to have sharp edges when not examined in too fine a detail. Figure 1.1 illustrates this situation for a point source S and a spherical body. The shadow on the screen S' is sharply defined. If S were an extended source (a source which when viewed from the point of observation S' subtends a finite angle), then the shadowing would have a more complex structure; but it would still be possible to analyze the shadow on the basis of rectilinear propagation of the light from each point on the source.

    Our treatment will be carried out in the paraxial-ray or Gaussian approximation. This approximation treats only those rays that lie close to

CHAPTER 1 INTRODUCTION

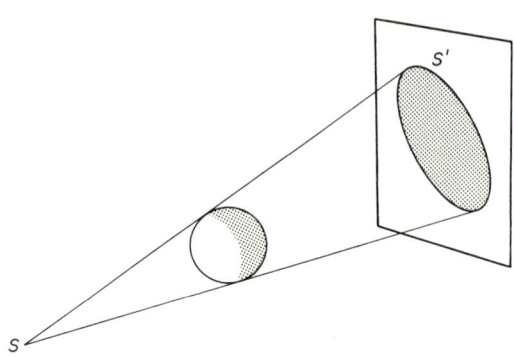

Fig. 1.1. Shadowing of the light from a point source S.

the axis of the system under examination and that make only small angles with this axis. While this may seem excessively restrictive, it turns out that the main properties of many optical systems are adequately described within the limits of this assumption. The details of the paraxial-ray assumption will be presented later in this chapter.

Fermat's Principle

Geometrical optics can be derived from a general principle known as Fermat's principle. This principle, which will lead to the laws of reflection and refraction (Snell's law), is stated as a variational principle similar to those which occur in other branches of physics. However, before we can state Fermat's principle, we must establish the idea of the optical path. Figure 1.2 shows the path (solid line) followed by a light ray passing

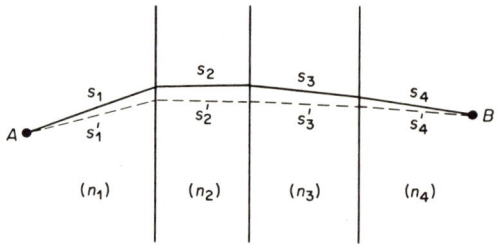

Fig. 1.2. Fermat construction showing the varied path.

from point A to point B through various media characterized by different refractive indices n_i. We assume that the surfaces as drawn separate the regions of different refractive index; thus, the path in any one region is a straight line. The optical path W from A to B is defined as

$$W = \sum_i n_i s_i, \qquad (1.1)$$

where the summation is taken over all the path segments. If the medium is one of continuously changing refractive index, the segment s_i is replaced by ds and W is then given by

$$W = \int_A^B n(s)\,ds, \qquad (1.1a)$$

where $n(s)$ indicates the position dependence of n.

We now turn our attention to a second path (dotted line) between A and B where the lengths of the path segments s_i' are only infinitesimally different from s_i. The path along the dotted line is then said to be a *variation* of the path along the solid line, and we write

$$W' = W + \delta W = \int n\,ds + \delta \int n\,ds. \qquad (1.2)$$

The δ here refers to a small variation in the total path and not to a single variable.

We can now state Fermat's principle:

If the path AB is the true optical path, the δW will vanish for small variations in the path.

In other words, the true optical path represents an extremum of W and thus a stationary value of the function W.

One must be careful in the application of Fermat's principle not to assume that the path is the minimum of W. One can easily construct cases where the true path W is the maximum, and often W is found to be only stationary in the sense of the inflection point of some three-dimensional surface. One important consequence of Fermat's principle is that, in the imaging of some object, all the optical paths are equal. We will see many examples of the use of this result throughout this book.

The index of refraction characterizing each medium is the ratio of the velocity of light in free space to the velocity of light within the medium. Since the sums or integrals involved in Fermat's principle involve products of distances and refractive indices, they are proportional to the times spent in the various media. An alternative statement of Fermat's principle is that the time of propagation of a light ray between two points is an extremum. Fermat's principle stated in this form will be found useful in some of the later discussions.

Reflection and Refraction

Figure 1.3 represents two possible ray paths from A to B where the ray is reflected at the mirror surface MM'. We will use Fermat's principle to distinguish between the true optical path and an arbitrarily constructed line joining A and B via some point on MM'.

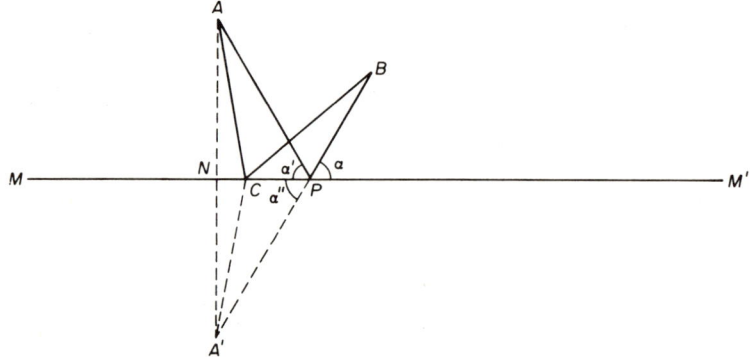

Fig. 1.3. Fermat construction for reflection.

We first construct ANA' normal to MM' so that AN and NA' are equal. From A' we construct $A'C$ and $A'P$ to the points where the ray is reflected at MM'. Since the triangles ANP and $A'NP$ are congruent, as are ANC and $A'NC$, the paths of the two rays APB and ACB can be replaced by $A'PB$ and $A'CB$, respectively. Since $A'PB$ is the straight-line path between A and B, it represents the minimum-length path that can be taken between A' and B, and in this case it represents the only extremum of the path since there is no maximum length path between these points. We can, therefore, conclude that the true path followed by the ray is APB and not ACB, since the latter does not satisfy the extremum condition of Fermat.

The geometry of the reflection shows that the angles α and α'' are equal, and from the congruence of ANP and $A'NP$ we have α'' equal to α' so that α and α' are equal. Figure 1.4 shows the details of the ray path at the point of reflection. We know that $\alpha=\alpha'$, and, by constructing the normal to the mirror surface at the point of reflection of the ray, we form two angles with the normal, which are also equal. These are the angle of

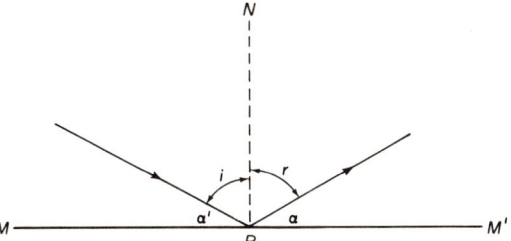

Fig. 1.4. Reflection showing the angle of incidence i and the angle of reflection r.

incidence, i, the angle made by the incident ray and the normal, and the angle of reflection, r, the angle made by the reflected ray and the normal. The law of reflection can then be stated:

The angle of incidence equals the angle of reflection.

It can also be shown that the incident ray, the normal to the reflecting surface at the point of reflection, and the reflected ray all lie in the same plane.

We now turn our attention to refraction of light. Figure 1.5 shows the

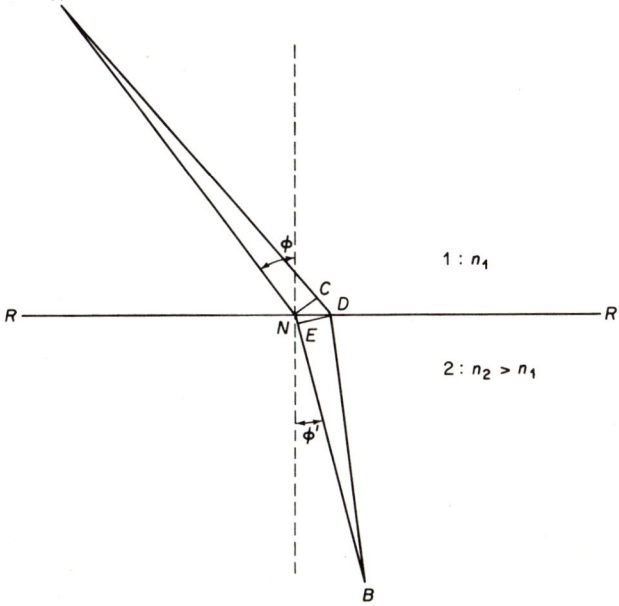

Fig. 1.5. Fermat construction for refraction.

Fermat construction for refraction. A ray passing through A is refracted at the surface RR' separating two media and then passes through point B. Medium 1 has index of refraction n_1, and medium 2 has an index of refraction n_2 which is greater than n_1. The index of refraction is simply defined, as in the previous section, as the ratio of the veclocity of light in free space (vacuum), c, to the velocity of light in the medium, v_m:

$$n = \frac{c}{v_m}. \tag{1.3}$$

At point N we construct the normal to the interface between 1 and 2. ANB is the experimentally observed path, and ADB is some other arbitrarily chosen path. Since we know the real path, which must satisfy Fermat's principle, we can use this principle to find the law for refraction. NC is normal to AD, and DE is normal to NB. We now consider the effect of moving point D closer and closer to point N. At the limit, since ANB is the actual path, the time of transit along CD must equal the time of transit along NE:

$$\frac{CD}{v_1} = \frac{NE}{v_2}. \tag{1.4}$$

If we multiply both sides of Eq. (1.4) by the velocity of light, c, and use Eq. (1.3), we find that

$$n_1[CD] = n_2[NE]. \tag{1.5}$$

We now note that angle CND is equal to ϕ and angle EDN is equal to ϕ' from the geometry of the right triangles NCE and NDC (at the limit). We then have $CD=[ND]\sin\phi$ and $NE=[ND]\sin\phi'$. Putting these into Eq. (1.5) and dividing by ND gives

$$n_1 \sin \phi = n_2 \sin \phi'. \tag{1.6}$$

This is Snell's law of refraction, due to W. Snell (1591–1626), who found this relationship experimentally. Since neither n_1 nor n_2 is specified, we can also write Snell's law in the form

$$n \sin \phi = \text{const}, \tag{1.7}$$

and this form is useful in cases where a number of different layers of media are involved. As is the case with reflected light, the incident ray, the normal at the point of refraction, and the refracted ray all lie in the same plane. Rays that pass into optically more dense regions (regions of higher index of refraction) are bent toward the normal, while those entering optically less dense regions are bent away from the normal.

We are dealing with optics in the Gaussian approximation in this book and thus with rays that make small angles with the interface between media. In the Gaussian approximation $\sin \phi \simeq \phi$, since the sine of an angle and the angle itself are approximately equal for small angles. In the Gaussian approximation, Snell's law is written

$$n\phi = \text{const}. \tag{1.8}$$

Refraction at a Spherical Surface

We next consider the situation illustrated in Fig. 1.6. A ray originating at a point P—the object point—at a distance p from a spherical refracting surface strikes the surface at point W and is bent back toward the axis POQ to strike the axis again at a distance q from the refracting surface. The axis of central symmetry of the system is generally known as the *optic axis*. We want to determine the relationship between p and q. All distances will be measured from the vertex O of the system. The *vertex* is defined as being the point of intersection of the refracting surface with the optic axis. Object distances—distances to the source of the optical signal and measured in the opposite direction to the incoming ray—will be considered positive; distances to the image will be considered to be positive if they are measured in the same direction as the outgoing ray. Radii of the refracting surfaces are positive if their center of curvature lies on the same side of their surfaces as the outgoing ray.

R is the radius of curvature of the refracting surface in Fig. 1.6. In this

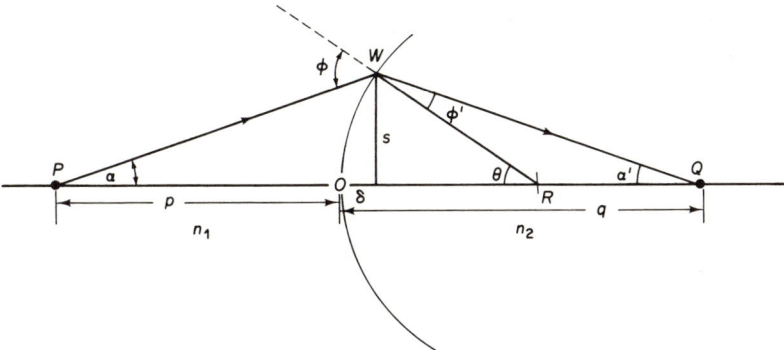

Fig. 1.6. Refraction at a spherical surface. The point P is the point source of the ray and Q is the point image.

case, applying the rule stated above, we have a positive radius of curvature. The radius RW, which is normal to the refracting surface, is extended in the figure so that ϕ and ϕ', the angle of incidence and the angle of refraction, respectively, can be easily located. Using the external angle theorem, we can write

$$\alpha + \theta = \phi \tag{1.9}$$

and

$$\theta = \alpha' + \phi' . \tag{1.10}$$

We are considering only rays that lie close to the optic axis (Gaussian approximation) so that Snell's law takes the form given in Eq. (1.8):

$$n_1 \phi = n_2 \phi' . \tag{1.11}$$

Combining Eqs. (1.9) and (1.11) gives

$$\phi' = \frac{n_1}{n_2}(\alpha + \theta), \tag{1.12}$$

which can be combined with Eq. (1.10) to eliminate ϕ' and gives

$$n_1 \alpha + n_2 \alpha' = (n_1 - n_2)\theta . \tag{1.13}$$

Equation (1.13) is not yet in a useful form since it contains the angle variables α, θ, and α' rather than the required distance variables p, q, and R. The tangents of the angles α, θ, and α' can be written:

$$\tan \alpha = \frac{s}{p - \delta}, \tag{1.14a}$$

$$\tan \alpha' = \frac{s}{q - \delta}, \tag{1.14b}$$

$$\tan \theta = \frac{s}{R - \delta}. \tag{1.14c}$$

In the Gaussian small-angle approximation, the tangents of the angles and the angles themselves are approximately equal. Also, the Gaussian approximation requires large radii of curvature for the refracting surfaces, i.e., paraxial rays, so that the small distances δ can be neglected. When this is done, Eqs. (1.14 a, b, c) become

$$\alpha = \frac{s}{p}, \tag{1.15a}$$

$$\alpha' = \frac{s}{q}, \tag{1.15b}$$

$$\theta = \frac{s}{R}, \tag{1.15c}$$

and these quantities may be inserted into Eq. (1.13) to give

$$\frac{n_1}{p} + \frac{n_2}{q} = \frac{n_2 - n_1}{R}. \tag{1.16}$$

This is the equation for refraction by a spherical surface. Note that since the equation does not involve the angles, *all rays* from p will pass through q, and q is then said to be the image point conjugate to the object (source) point p. When some of the rays do not satisfy the Gaussian approximation, the image is spread out along the optic axis near p rather than falling strictly at the point p. This is called an *image aberration*; aberrations are dealt with later in this book.

Equation (1.16) is valid for concave surfaces as well as for convex surfaces, but we will not go through the proof here. The image in Fig. 1.6 is *real* since the light rays actually pass through the point q. It is essential to realize, however, that in some cases the image distance q obtained in the solution of Eq. (1.16) will be negative. This implies that the image is formed on the same side of the surface as the object. Such an image is called *virtual*. In Fig. 1.7 we illustrate this situation. Rays originating at

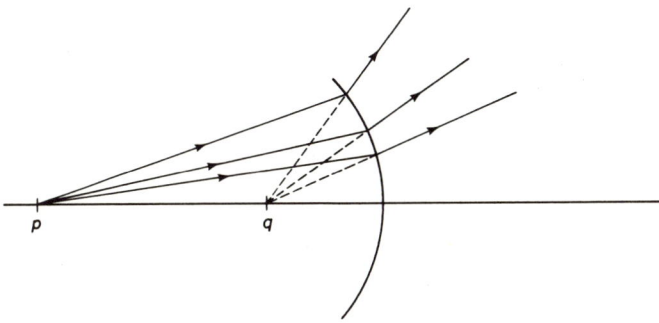

Fig. 1.7. Formation of a virtual image by a surface with a concave refracting surface.

p are bent away from the optic axis. To an observer on the right of the refracting surface the rays appear to originate at q. The image in this case is virtual since all rays appear to converge; however, they do not pass through the point of convergence.

Similarly, an object may be virtual if it is not the source of the rays, but only the apparent source of the rays.

The Thin Lens

The thin lens is formed by two refracting surfaces in close proximity. The assumption which applies to the thin lens is that the two refracting surfaces that form the lens lie at the same point along the optic axis. In the Gaussian approximation this is a reasonable assumption, particularly where the radii of the spherical surfaces making up the lens are large. Various forms of the thin lens are illustrated in Fig. 1.8.

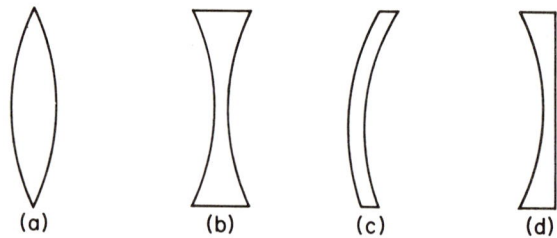

Fig. 1.8. Four forms of thin lenses. (a) Double-convex. (b) Double-concave. (c) Convexo-concave. (d) Plano-concave.

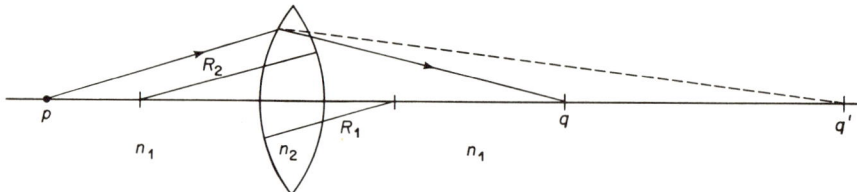

Fig. 1.9. Image formation by a thin lens.

Figure 1.9 illustrates an object-image set of conjugate points for a double convex thin lens. We can use Eq. (1.16) to derive the relationship between p and q. The index of refraction of the lens is taken as n_2 and the index of the medium as n_1. The radii of curvature of the surfaces are R_1 and R_2, respectively.

Consider the effect of the first surface on the incoming ray. Equation (1.16) gives

$$\frac{n_1}{p} + \frac{n_2}{q'} = \frac{(n_2 - n_1)}{R_1}. \tag{1.17}$$

This gives us the effect of the first surface. The rays striking the second surface will appear to come from q' and will be acted on by the surface R_2. According to our sign convention, q' will be negative and, again, using Eq. (1.16), we get

$$\frac{n_2}{-q'} + \frac{n_1}{q} = \frac{(n_1 - n_2)}{R_2}. \qquad (1.18)$$

Adding Eqs. (1.17) and (1.18) gives the thin lens equation:

$$\frac{n_1}{p} + \frac{n_2}{q} = (n_2 - n_1)\left(\frac{1}{R_1} - \frac{1}{R_2}\right). \qquad (1.19)$$

This is the usual reciprocal law for lenses, and in the case where the lens is in air ($n_1 = 1$), the equation becomes

$$\frac{1}{p} + \frac{1}{q} = (n-1)\left(\frac{1}{R_1} - \frac{1}{R_2}\right). \qquad (1.20)$$

The quantity on the right-hand side of this equation is often equated to $1/f$, the reciprocal of the focal length, so that

$$\frac{1}{f} = (n-1)\left(\frac{1}{R_1} - \frac{1}{R_2}\right). \qquad (1.21)$$

This is known as the lensmaker's equation. The *focal length* is the distance from the lens at which an object at infinity is imaged (first focal length), or the distance from the lens at which an object must be placed to form an image at infinity (second focal length). In the case of the thin lens, the first and second focal lengths are equidistant from the lens and are located on opposite sides of the lens.

We will not actually operate with the lens equation in this form but will derive in subsequent chapters a more general way of treating lenses. We are, however, required to show at some point that both the focal length as given in Eq. (1.21) and the lens equation (1.20), as derived here, are consistent with this new treatment.

Images Formed by Reflection

In order to complete this brief introduction to the algebraic formulation of geometric optics, we now turn to the derivation of the object-image relationships for a spherical reflecting surface. We again apply the Gaussian assumption, i.e., the reflecting surfaces have large radii of curvature and all rays make small angles with the optic axis.

Figure 1.10 illustrates the situation. The incoming ray strikes the mirror surface at W and is reflected. The incident and reflected rays make

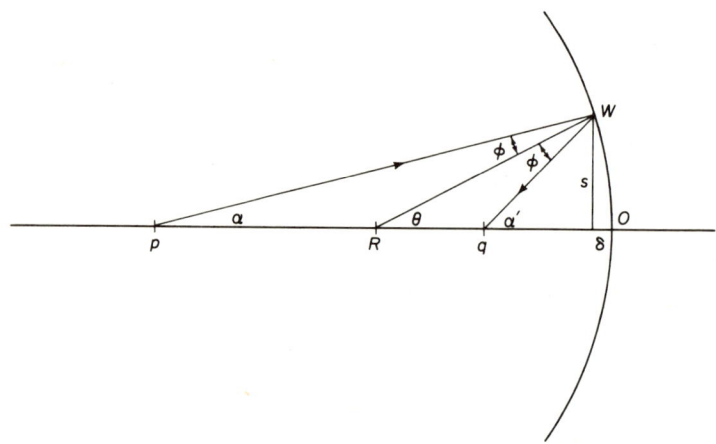

Fig. 1.10. Image formation by a spherical reflecting surface.

equal angles with the normal to the mirror surface at W, the normal being the radius vector to W. We proceed as in the lens derivation by forming angle equalities. Thus, from the exterior angle theorem, we can write

$$\theta = \phi + \alpha \tag{1.22a}$$

and

$$\alpha' = \theta + \phi. \tag{1.22b}$$

Eliminating ϕ we get

$$\alpha + \alpha' = 2\theta. \tag{1.23}$$

Once again, we have an equation involving angles, and it is necessary to replace the angle variables with distance variables. We note that in the Gaussian approximation

$$\tan \alpha \simeq \alpha = \frac{s}{p-\delta}, \tag{1.24a}$$

$$\tan \alpha' \simeq \alpha' = \frac{s}{q-\delta}, \tag{1.24b}$$

$$\tan \theta \simeq \theta = \frac{s}{R-\delta}. \tag{1.24c}$$

As with lenses, δ is a small quantity in the Gaussian approximation and we can neglect it here. This will again lead to aberrations, but the effect will be minor. Substituting in Eq. (1.23) with (1.24a) to (1.24c) we get

$$\frac{1}{p} + \frac{1}{q} = \frac{2}{R}, \tag{1.25}$$

and this is the mirror equation. The focal length is given by

$$\frac{1}{f} = \frac{2}{R} \qquad (1.26)$$

and again represents the point conjugate with infinity.

We have assumed throughout this derivation that the radius of curvature was positive. For concave mirrors, the radius of curvature is always taken as positive here and for convex mirrors the radius is negative. We note that virtual images can be formed when we deal with convex mirrors. The same considerations apply here to virtual images as for lenses, and additionally, we find for virtual mirror images that the image is formed behind the mirror.

The plane mirror is an interesting example of the formation of virtual images. For a plane mirror $R = \infty$, and the mirror equation (1.25) reduces to

$$\frac{1}{p} = -\frac{1}{q} \qquad (1.27)$$

or $p = -q$. Thus with plane mirrors the image is always virtual (q negative) and the image appears to be at a distance behind the mirror equal to the distance at which the object is placed in front of the mirror.

Problems

1.1 Consider an ellipsoidal mirror with a major axis of 10 cm and a minor axis of 6 cm. The foci are located 4 cm from the center. Only two rays pass through the center from a source at one focus to a detector at the other. Discuss the imaging of the source in terms of Fermat's principle and in particular discuss the rays passing through the center.

1.2 A submarine has a window made of 12-in. thick glass, $n_{H_2O} = 1.333$. A sailor sees a fish along a line that makes an angle of 60° with the normal to the window. How far from the extension of this line is the fish if it is 1 ft from the window?

1.3 A still lake is covered by 2 cm of oil, index 1.635. A ray originates in the water and strikes the water-oil interface at an angle of 45°. Will this ray appear in the air above the oil?

1.4 A prism is made in the following form:

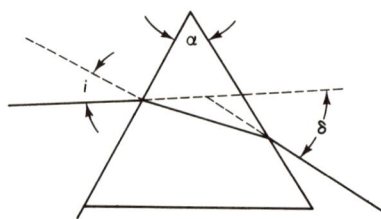

Find the angle of deviation δ in terms of the incident angle and the apex angle α for a prism with index n. Investigate the limiting case of small α and small incident angles.

1.5 An object is embedded 1 cm below the surface of a glass marble with index 1.6 and a radius of 2 cm. Find the apparent position of this object as seen by someone looking into the marble from either side along the diameter including the object.

1.6 A parallel beam of light is projected on a clear glass marble 2 cm in diameter with index 1.5. Where does the beam converge?

1.7 A thin glass lens has index 1.6 and a focal length of 10 cm. Does this uniquely specify the shape of the lens? Will the statement that the lens has a symmetric shape help?

1.8 What effect will changing the external medium have on the properties of a mirror?

1.9 A man uses a concave mirror with a radius of 50 cm for shaving. If his face is 15 cm in front of the mirror, where is the image formed? Is the image larger or smaller, and would this be effective for shaving?

1.10 A small reflecting telescope has a mirror with a 2-m radius of curvature. Where is the focal plane relative to the mirror, and is the mirror ground concave or convex?

Chapter 2

Refraction and Translation Matrices

In the preceding chapter the thin lens equation was developed from the study of the effect of spherical refracting surfaces on a light ray. In this chapter we develop the basis of the matrix formalism of geometric optics. Subsequently, we show this formalism to be equivalent to that developed in Chapter 1. The advantage of the matrix formalism rests in the fact that it lends itself to an easy treatment of completely general optical systems, as well as to special cases such as the thin lens, within one formal framework.

Sign Conventions

In order to achieve uniformity in notation we must initially agree on the sign convention. At this point only those conventions needed here are introduced, later other rules are added as required. The number of rules is minimal, but each must be mastered in turn so that calculations can be carried out.

In general, and unless otherwise stated, rays will be taken to move

16   CHAPTER 2   REFRACTION AND TRANSLATION MATRICES

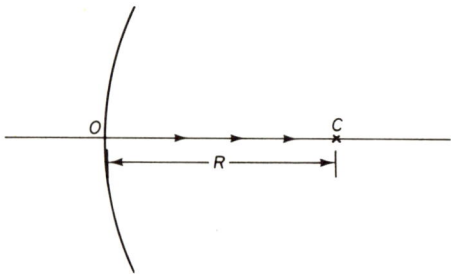

Fig. 2.1. A spherical surface with positive radius of curvature, $+R$. The radius is measured from the vertex O of the surface to the center of curvature C from left to right as indicated by the arrows on the optic axis.

from left to right, and distances when measured from left to right will be thought of as positive. These distances will always be measured along the optic axis from the vertices; that is, from the point of intersection of a refracting surface and the optic axis. This means, of course, that the radius of curvature of a convex surface (Fig. 2.1) is positive.

Displacements normal to the optic axis, i.e., along the y axis, will follow the usual usage, with displacements above the axis positive and those below negative.

The Refraction Matrix

We will now determine the effect of a spherical refracting surface on a ray striking the surface. This procedure is similar to that which was carried out in Chapter 1 for a spherical surface, but here the variables will be chosen differently. We take as variables the height y at which the incoming ray strikes the refracting surface and l, the cosine of the angle made by the ray with the y axis. These are the only variables required so long as the system may be thought of as being axially symmetric about the optic axis. Axial symmetry is a fundamental assumption of this part of the theory.

Figure 2.2 illustrates the cross section of the optical system taken so as to include the optic axis. We now examine the ray ABD moving from A to B in a medium of index n_1 and then moving forward toward D in a medium of index n_2. BN represents an extension of AB in region 2, while BD represents the actual path in region 2. The radius of curvature R of the spherical surface is given by BC.

We proceed by making a construction commonly used in geometric

THE REFRACTION MATRIX

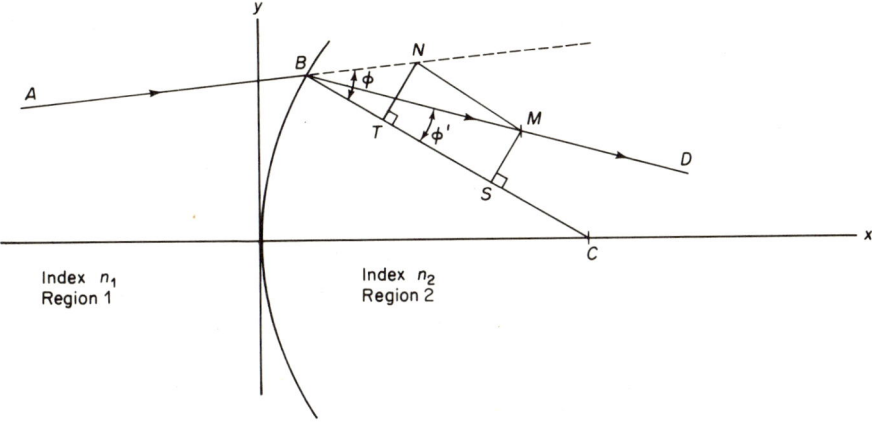

Fig. 2.2. Refraction at a spherical surface. The ray AB undergoes an abrupt change in direction as a result of the change of refractive index.

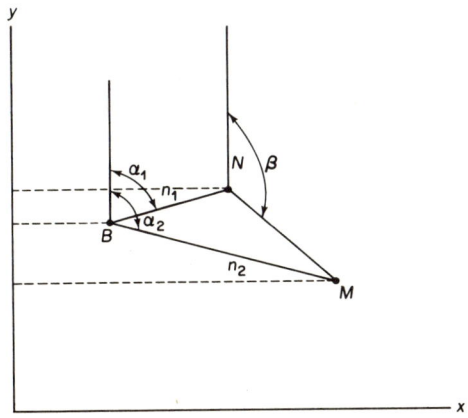

Fig. 2.3. Details of the triangle BNM from Fig. 2.2 showing the angles involved in the definition of the direction cosines.

optics. Along the extension BN of AB we mark off a length proportional to the index n_1 and drop a normal from this point to BC. The length BN is then equal to kn_1, where k is the proportionality constant; k merely converts the dimensionless quantity n into a length and we will take $k = 1$. Along BD, the actual ray, we lay off a length kn_2 and again drop a perpendicular to BC. The length $NT = BN \sin \phi = kn_1 \sin \phi$, and likewise $MS = BM \sin \phi' = kn_2 \sin \phi'$. Since by Snell's law $n_1 \sin \phi = n_2 \sin \phi'$ and $kn_1 \sin \phi = kn_2 \sin \phi'$, we can conclude that $NT = MS$. This immediately establishes NM as parallel to BC, and the angle β, Fig. 2.3, which NM

makes with the positive y axis is the same as that made with the y axis by BC. We can immediately evaluate $\cos \beta$ from the geometry as

$$\cos \beta = \frac{0 - y_B}{R} = -\frac{y_B}{R}, \tag{2.1}$$

where y_B is the y coordinate of point B.

We now turn our attention to the triangle BNM which we have redrawn in Fig. 2.3. We are now going to project the sides of this triangle onto the y axis. BN makes the same angle with the positive y axis as does the original ray AB, and the cosine of this angle α_1 (which is in fact a direction cosine) is l_1. The cosine of the angle α_2 between the positive y axis and BM we call l_2. The projection of NM on the y axis is the sum of the projections of BN and BM, and we can write

$$NM \cos(\pi - \beta) = n_1 \cos \alpha_1 + n_2 \cos(\pi - \alpha_2).$$

This expression can be rewritten using the usual trigonometric identities as

$$-NM \cos \beta = n_1 \cos \alpha_1 - n_2 \cos \alpha_2$$

where $\cos \alpha_1 = l_1$ and $\cos \alpha_2 = l_2$ from the definition above and

$$n_2 l_2 = n_1 l_1 + NM \cos \beta.$$

We have already found $\cos \beta$ in Eq. (2.1), and thus

$$n_2 l_2 = n_1 l_1 - NM \frac{y_B}{R}. \tag{2.2}$$

This gives the direction cosine of the refracted ray in terms of the direction cosine of the incident ray and the height at which the incident ray strikes the refracting surface. The quantity NM, however, must be eliminated from (2.2) if the expression is to be at all general. If we refer to Fig. 2.2 once more and note that $NM = TS$ and that

$$TS = n_2 \cos \phi' - n_1 \cos \phi,$$

we can rewrite Eq. (2.2) in the form

$$n_2 l_2 = n_1 l_1 - \frac{n_2 \cos \phi' - n_1 \cos \phi}{R} y_B \tag{2.3}$$

or

$$n_2 l_2 = n_1 l_1 - P y_B \tag{2.3a}$$

where

$$P = \frac{n_2 \cos \phi' - n_1 \cos \phi}{R} \qquad (2.4)$$

is called the *power of the surface*.

For convenience we choose to define new direction cosines by weighting the true direction cosines by the index of the medium so that

$$n_2 l_2 = \Lambda_2$$
$$n_1 l_1 = \Lambda_1$$

and Eq. (2.3a) now becomes

$$\Lambda_2 = \Lambda_1 - P y_B. \qquad (2.5)$$

The effect on a ray of an abrupt change in index of refraction merely changes the direction of the ray but in no way affects the height of the ray at the surface. We can write

$$y_{B1} = y_{B2}, \qquad (2.6)$$

where y_{B1} is the distance measured from the optic axis to the point of intersection of the ray and the refracting surface in region 1, and y_{B2} is similarly defined in region 2.

The pair of Eqs. (2.5) and (2.6) can conveniently be expressed in a matrix equation:

$$\begin{pmatrix} \Lambda_2 \\ y_{B2} \end{pmatrix} = \begin{pmatrix} 1 & -P \\ 0 & 1 \end{pmatrix} \begin{pmatrix} \Lambda_1 \\ y_{B1} \end{pmatrix}, \qquad (2.7)$$

where

$$\mathscr{R} = \begin{pmatrix} 1 & -P \\ 0 & 1 \end{pmatrix}$$

is called the *refraction matrix*.

The column vectors

$$\begin{pmatrix} \Lambda_1 \\ y_{B1} \end{pmatrix}, \begin{pmatrix} \Lambda_2 \\ y_{B2} \end{pmatrix}$$

are the parameters which we use to characterize the ray in regions 1 and 2, respectively. $\mathscr{R}$ is then a transformation which carries the incident ray in medium 1 into the refracted ray in medium 2.

Translation Matrix

In addition to the refraction of a ray which leads to abrupt changes in the direction of the ray, a second process occurs in optical systems.

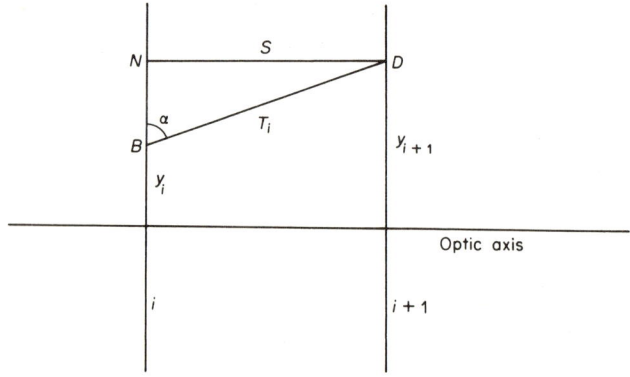

Fig. 2.4. Geometry of a ray in the uniform medium of index n_i between two arbitrary surfaces.

This process is the translation of the ray between the refractive (or as we shall see later—reflective) surfaces. Figure 2.4 illustrates the situation.

The ray BD moves in a uniform medium of index n_i between the arbitrary surfaces i and $i+1$. In practice these surfaces will usually be refracting surfaces, but we are free to be perfectly arbitrary in defining them. Since the medium is uniform the ray will move in a straight-line path from B to D and the angle with the y axis will not change since Λ_i and Λ_{i+1} are in the same medium. We can then immediately write

$$\Lambda_i = \Lambda_{i+1} \tag{2.8}$$

and our only remaining problem is the determination of y_{i+1} relative to y_i. From Fig. 2.4 we see that

$$y_{i+1} = y_i + T_i \cos \alpha, \tag{2.9}$$

where T_i is the path length in medium i. Cos α can be replaced by the equivalent expression Λ_i/n_i, in which case Eq. (2.9) becomes

$$y_{i+1} = y_i + \frac{T_i}{n_i}\Lambda_i. \tag{2.10}$$

As in the case of the refraction matrix, we replace the translation distance T_i by $\mathbf{T}_i' = T_i/n_i$. $\mathbf{T}_i'$ is the "reduced optical length" and is merely the true path length in medium i weighted in this case by the reciprocal of the refractive index or $1/n_i$.

Equations (2.8) and (2.10) may be written in matrix form using the same vectors as were used for the refraction matrix (2.7), to give

$$\begin{pmatrix} \Lambda_{i+1} \\ y_{i+1} \end{pmatrix} = \begin{pmatrix} 1 & 0 \\ \mathbf{T}_i' & 1 \end{pmatrix}\begin{pmatrix} \Lambda_i \\ y_i \end{pmatrix}, \tag{2.11}$$

where

$$(\mathcal{T}_i') = \begin{pmatrix} 1 & 0 \\ \mathbf{T}_{i'} & 1 \end{pmatrix}$$

is the *translation matrix*.

Again we have a transformation in the form of a 2×2 matrix which generates the ray properties at the $i+1$ surface from those at the i surface.

The Paraxial-Ray Assumption

We have now developed two matrices, $(\mathcal{R})$ and $(\mathcal{T})$, which can act as transformations on the column vectors describing a light ray. These matrices are particularly simple in form, being 2×2 and each having only one of its elements different from zero or unity. However, the structure of the remaining elements is somewhat complex. In both cases the structure depends on the optical direction cosine of the incoming ray. It would obviously be advantageous to eliminate from these matrix elements those factors that depend on the incident ray and thereby make the matrices $(\mathcal{R})$ and $(\mathcal{T})$ constants characteristic of the particular surface or segment of the particular optical system being examined.

We have already dealt with an approximation in Chapter 1 (page 8) that allowed us to remove a term involving a trigonometric function and to replace it with a simple ratio. We will again make the same Gaussian assumption; namely, we will assume that all rays are confined to a cylinder close to the optic axis. This means that in the case of the translation matrix $(\mathcal{T})$ the element T_i, which represents the distance BD in Fig. 2.4, can be replaced by $S = ND$, the axial separation of the surfaces, since the angle α must be close to $\pi/2$ radians.

Similarly, the matrix element P, which appears in the refraction matrix $(\mathcal{R})$ and which has the form given in Eq. (2.4):

$$P = \frac{n_2 \cos\phi' - n_1 \cos\phi}{R}, \qquad (2.4)$$

can be simplified. In the paraxial-ray assumption both ϕ and ϕ' must be close to zero radians since they are restricted to be nearly parallel to the optic axis. We then assume that

$$\cos\phi' \approx 1 \quad \text{and} \quad \cos\phi \approx 1,$$

so that P now takes the form

$$P \approx p = \frac{n_2 - n_1}{R}, \tag{2.12}$$

and p as defined is independent of the angle at which the ray strikes or leaves the spherical refracting surface (Fig. 2.2) and is therefore a constant which characterizes the surface in terms of the change in refractive index across the surface and the radius of curvature of the surface.

We now replace the matrices $(\mathscr{T})$ and $(\mathscr{R})$ by their form in the paraxial-ray limit. We will call these new matrices (ℓ) and (ι), respectively. In general, we will use lower-case script letters to represent matrices in the paraxial-ray limit. $(\mathscr{R})$ now becomes

$$(\iota) = \begin{pmatrix} 1 & -\dfrac{n_2 - n_1}{R} \\ 0 & 1 \end{pmatrix} = \begin{pmatrix} 1 & -p \\ 0 & 1 \end{pmatrix}. \tag{2.13}$$

and $(\mathscr{T})$ becomes

$$(\ell) = \begin{pmatrix} 1 & 0 \\ \dfrac{s}{n} & 1 \end{pmatrix} = \begin{pmatrix} 1 & 0 \\ \mathbf{t} & 0 \end{pmatrix} \tag{2.14}$$

and both (ι) and (ℓ) are constants characteristic of the system being studied.

The paraxial-ray assumption also produces a simplification in the ray vectors in a way that is not obvious. If we return to Fig. 2.2, we see that y_B, the height parameter of the ray, is different by a small amount from the height at which the ray crosses the y axis. This difference in height varies with both the height of the ray, y_B, and the angle at which it strikes

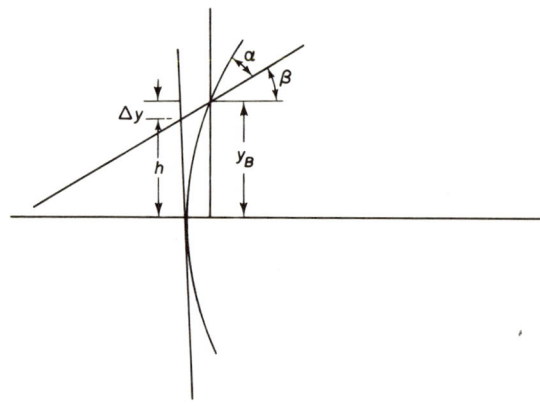

Fig. 2.5. Parameters involved in the paraxial-ray assumption.

the refracting surface. A ray that strikes the surface at the optic axis, i.e., the vertex of the refracting surface, has a Δy, as defined in Fig. 2.5, equal to zero regardless of the angle at which it strikes the surface. On the other hand, a ray parallel to the optic axis will have Δy zero regardless of its height above the axis.

The paraxial-ray assumption takes $\Delta y \approx 0$ for all rays, and therefore in the ray vector we replace y_B with h, the displacement relative to the optic axis.

Λ, the optical direction cosine, is given by

$$\Lambda_1 = n_1 \cos \alpha_1$$

and

$$\Lambda_2 = n_2 \cos \alpha_2 ,$$

so that the optical direction cosines are nonlinear in the sense that they depend on the cosines of the angles rather than on the angles themselves. In the paraxial-ray assumption, $\cos \alpha = \sin(90-\alpha) \approx \beta$ for α near $\pi/2$, and we replace Λ by $\lambda = n\beta$ in the expression for the ray vector. Ray vectors are then taken to be of the form

$$\begin{pmatrix} \lambda \\ h \end{pmatrix}$$

in the paraxial-ray limit.

It should be clear at this point that the paraxial-ray assumption as treated here is the same assumption that was applied in Chapter 1 in deriving the expression for the conjugate points of a spherical refracting surface. Replacing y_B with h is equivalent to neglecting the distance δ or, in other words, to the assumption that the radius of curvature of the refracting surface is large compared to the other parameters of the system. The replacement of $\sin \beta$ with β is precisely the assumption made in the earlier case.

It is instructive to look at the effect of a plane interface on a ray at this point since it illustrates the effect of the assumptions we have made. For a plane surface, the radius of curvature is infinite and the power of the surface is given by

$$p = \lim_{R \to \infty} \frac{n_2 - n_1}{R} = 0 ,$$

so that the refraction matrix

$$(\ell) = \begin{pmatrix} 1 & 0 \\ 0 & 1 \end{pmatrix}$$

is the unit matrix.

Our matrix equation within the paraxial-ray assumption is then

$$\begin{pmatrix} \lambda_2 \\ h_2 \end{pmatrix} = \begin{pmatrix} 1 & 0 \\ 0 & 1 \end{pmatrix} \begin{pmatrix} \lambda_1 \\ h_1 \end{pmatrix},$$

and we have the results

$$h_2 = h_1$$

and

$$\lambda_2 = \lambda_1.$$

The first of these is trivial but the second can be written with the λ's replaced by their equivalent $n\beta$, and we get (See Fig. 2.6),

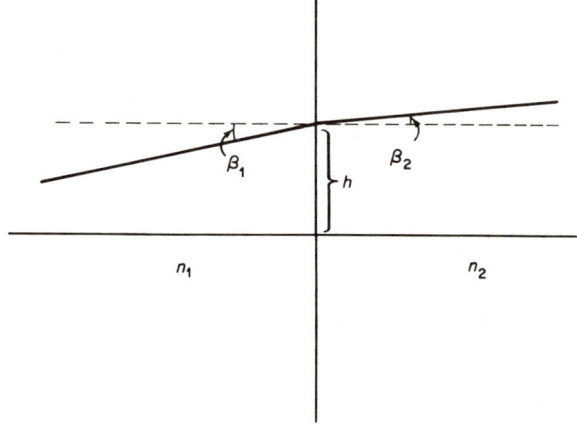

Fig. 2.6. Refraction at a spherical surface of infinite radius of curvature in the paraxial-ray assumption.

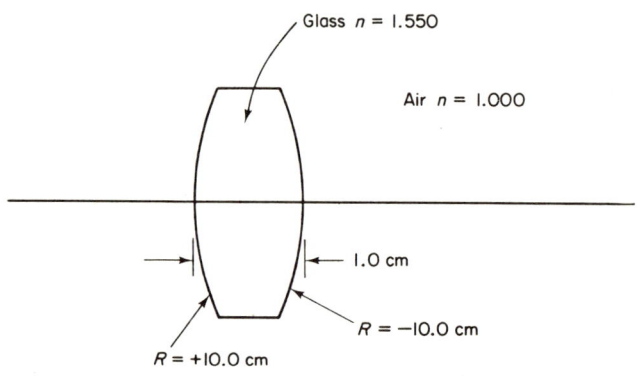

Fig. 2.7. A typical thick lens.

$$n_1\beta_1 = n_2\beta_2.$$

This equation is identical with Snell's law in the paraxial limit Eq. (1.8). The consistency of this approach is thus illustrated.

As an example we can consider the thick lens illustrated in Fig. 2.7. We assume that the lens is in air ($n=1.000$) and that the lens is constructed from glass $n=1.5500$. The radii of curvature of both surfaces is taken as 10 cm and the lens is 1 cm thick. There are three matrices associated with this lens, two refraction matrices and one translation matrix. We will construct them in turn.

The first lens surface r_1 in Fig. 2.7 has a positive radius of curvature since the center of curvature of this surface lies to the right of the vertex A. The power p of this surface is given as in Eq. (2.13) as

$$p_1 = \frac{n_2 - n_1}{R} = \frac{1.5500 - 1.0000}{10} = 0.0550$$

and the refraction matrix (ℓ_1) is

$$(\ell_1) = \begin{pmatrix} 1 & -0.0550 \\ 0 & 1 \end{pmatrix}.$$

For ℓ_2 the power is given by:

$$p_2 = \frac{n_1 - n_2}{R} = \frac{1.0000 - 1.5500}{-10} = 0.0550.$$

Note that $n_1 - n_2$ is negative in this case since the index n_1, that is, the index on the right side of the surface is less than the index n_2 on the left side. R is also negative since the center of curvature lies to the left of the vertex B. As a result p_1 and p_2 are the same and

$$(\ell_2) = (\ell_1) = \begin{pmatrix} 1 & -0.0550 \\ 0 & 1 \end{pmatrix}.$$

Finally, the translation matrix $(\mathscr{T})$ is found by first computing the reduced optical length of the lens, given by

$$\frac{s}{n} = \frac{1.0000}{1.5500} = 0.645$$

and

$$(\mathscr{T}) = \begin{pmatrix} 1 & 0 \\ 0.645 & 1 \end{pmatrix}.$$

In the ensuing chapters the matrices developed here will be applied to various optical systems in order to extend the ideas just introduced.

Problems

2.1 Make a diagram that shows the various sign conventions used here.

2.2 Plot the power of a surface $R=50$ cm separating air from glass $n=1.50$ as a function of the incident angle ϕ. Why is this of interest?

2.3 The column vectors appearing in Eqs. (1.17) and (1.11) can also be written as row vectors. What form will these equations take under those circumstances?

2.4 Plot the reduced optical length as a function of angle for a 10-cm linear displacement in several different media.

2.5 Relate the paraxial-ray assumption to the graphs obtained in Problems 2.2 and 2.4.

2.6 Write the equations for the paraxial-ray theory in terms of row matrices.

2.7 An equiconcave lens in air has radii of 15 cm and a thickness of 3 cm. If $n=1.5$, find the system matrix.

2.8 The lens of Problem 2.7 is placed at the side of a tank containing oil with $n=1.4$. How does the system matrix change?

Chapter 3

The Optical System; Imaging

In the previous chapter we developed a set of equations which allowed us to trace the path of light through a refracting element. In addition it was shown that both the refraction at each refracting surface and the transit of the ray between surfaces could be represented by a system of matrices. We now want to examine the situation where a number of refracting surfaces and the accompanying translations are combined to form an optical system. This is illustrated in Fig. 3.1. We set aside for the time being the question of reflecting surfaces.

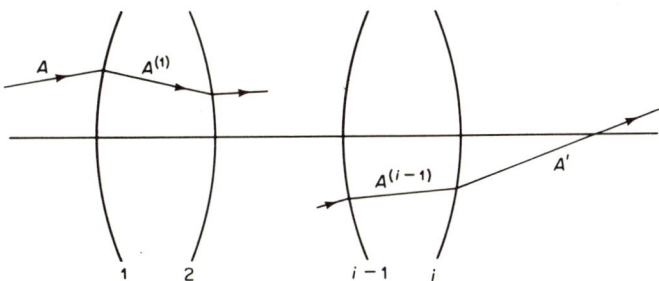

Fig. 3.1. A general optical system with i refracting surfaces.

28   CHAPTER 3   THE OPTICAL SYSTEM; IMAGING

In Fig. 3.1 we have a ray A entering the system on the left through refracting surface 1. This ray changes direction at surface 1, is then translated to surface 2 where it undergoes another change in direction, and so on until it reappears as ray A' to the right of the optical system. The system may be thought of as transforming the ray A from the space to the left of the system (object space) into the ray A' in the space to the right of the system (image space). We have already developed the elements which we can use to handle the required mathematical ray tracing. We characterize the incoming ray A by its ray vector:

$$\begin{pmatrix} \lambda \\ h \end{pmatrix}_A. \qquad (3.1)$$

The effect of surface 1 is to transform the ray vector (1) into a new ray vector $A^{(1)}$. This is done using the refraction matrix for 1 in the paraxial-ray assumption and we have a new ray vector $A^{(1)}$ given by

$$\begin{pmatrix} \lambda \\ h \end{pmatrix}_{A^{(1)}} = \begin{pmatrix} 1 & -p_1 \\ 0 & 1 \end{pmatrix} \begin{pmatrix} \lambda \\ h \end{pmatrix}_A. \qquad (3.2)$$

$A^{(1)}$ is then translated to surface 2 and the translation can be accounted for by operating on the ray vector $A^{(1)}$ with a translation matrix (Eq. 2.14),*

$$\begin{pmatrix} \lambda \\ h \end{pmatrix}_{A_t^{(1)}} = \begin{pmatrix} 1 & 0 \\ \dfrac{s_{1,2}}{n_{1,2}} & 1 \end{pmatrix} \begin{pmatrix} \lambda \\ h \end{pmatrix}_{A^{(1)}}$$

$$= \begin{pmatrix} 1 & 0 \\ \dfrac{s_{1,2}}{n_{1,2}} & 1 \end{pmatrix} \begin{pmatrix} 1 & -p_1 \\ 0 & 1 \end{pmatrix} \begin{pmatrix} \lambda \\ h \end{pmatrix}_A.$$

Eventually, by continuing this process, we can find the ray vector A' by operating on the ray vector until we have taken account of the i refracting surfaces and the $(i-1)$ translations associated with the optical system. We then get an equation in the form

$$\begin{pmatrix} \lambda \\ h \end{pmatrix}_{A'} = \begin{pmatrix} 1 & -p_i \\ 0 & 1 \end{pmatrix} \begin{pmatrix} 1 & 0 \\ \dfrac{s_{(i-1),i}}{n_{(i-1),i}} & 1 \end{pmatrix} \cdots \begin{pmatrix} 1 & 0 \\ \dfrac{s_{1,2}}{n_{1,2}} & 1 \end{pmatrix} \begin{pmatrix} 1 & -p_1 \\ 0 & 1 \end{pmatrix} \begin{pmatrix} \lambda \\ h \end{pmatrix}_A. \qquad (3.3)$$

Equation (3.3) may be reduced by having each matrix operate in turn

---

* By $A_t^{(1)}$ we mean the vector $A^{(1)}$ translated to the next optical surface, where $A^{(1)}$ is the ray vector at surface 1 after refraction. $n_{1,2}$ is the index of refraction of the region between surfaces 1 and 2.

on the ray vector at the right using the standard rules for matrix multiplication, and eventually on either side of Eq. (3.3) we will have ray vectors, and the equation will be solved. We will choose instead to combine the matrices so that we get a single matrix which characterizes the entire system. This matrix, the product of the $(2i-1)$ two-by-two matrices in Eq. (3.3), is known as the *system matrix*, and this system matrix will contain all the information of interest within the limit of the paraxial-ray assumption.

One of the simplest optical systems is, of course, just a single lens, as illustrated in Fig. 2.6. We have already found all the matrices associated with this system, viz., the two refraction matrices and the translation matrix. The system matrix will then be given by $(\ell_2)(\prime)(\ell_1)$ where, we point out again, the ordering of the matrices is just the opposite of the ordering of the elements of the system. The system matrix is then given by

$$\begin{pmatrix} 1 & -p_2 \\ 0 & 1 \end{pmatrix} \begin{pmatrix} 1 & 0 \\ \frac{s_{1,2}}{n_{1,2}} & 1 \end{pmatrix} \begin{pmatrix} 1 & -p_1 \\ 0 & 1 \end{pmatrix}$$

$$= \begin{pmatrix} 1 & -0.055 \\ 0 & 1 \end{pmatrix} \begin{pmatrix} 1 & 0 \\ 0.645 & 1 \end{pmatrix} \begin{pmatrix} 1 & -0.055 \\ 0 & 1 \end{pmatrix}$$

$$= \begin{pmatrix} 0.96452 & -0.10805 \\ 0.645 & 0.96452 \end{pmatrix},$$

and we see that while the refraction and translation matrices have elements 0 and 1 in three of their four positions, the system matrix generally has nonintegral elements at all positions. The general system matrix is taken in the form

$$\begin{pmatrix} b & -a \\ -d & c \end{pmatrix}, \qquad (3.4)$$

where the lower case letters are used to denote the matrix formed under the paraxial-ray assumption. The negative signs on the elements a and d are introduced at this point for later convenience.

An extremely useful device is available for checking system matrices once they have been computed. Since the system matrix generally contains four nonintegral elements, we find it convenient to apply a check based on the theorem from linear algebra which states that the determinant of a matrix formed as the product of two or more matrices is equal to the product of the determinants of the component matrices.*
This theorem can be applied to any matrix product where the deter-

---

* See Appendix 1.

minants exist, but in this case the special form of the translation matrix is unity:

$$\begin{vmatrix} 1 & 0 \\ \dfrac{s}{n} & 1 \end{vmatrix} = 1,$$

and in the same way the determinant of any refraction matrix also equals one:

$$\begin{vmatrix} 1 & -p \\ 0 & 1 \end{vmatrix} = 1.$$

Since any system matrix is a product of only refraction and translation matrices, we can state:

the determinant of the system matrix is unity,

and a simple condition which each system matrix must meet is established.

If we apply this rule to the simple system for which we just found the system matrix, we have

$$\begin{vmatrix} 0.96452 & -0.10805 \\ 0.645 & 0.96452 \end{vmatrix} = 0.99999 \approx 1.$$

and we see that our optical matrix meets this condition for this simple system. The determinant is not *exactly* unity since approximations have been made in expressing its elements as decimals. It should be clear that this condition of a unit determinant is a necessary but not a sufficient condition on the system matrix. If we arbitrarily write a 2×2 matrix with unit determinant, it need not necessarily correspond to a real physical situation.

We can apply this test at any stage in the computation of the system matrix, and in cases where several lenses are used in combination to form an optical system it is particularly useful to apply this test frequently, i.e., to test at various stages in the computation to insure that the determinant is equal to one.

Imaging

We now turn to the problem of image formation. Figure 3.2 illustrates the geometry of the problem. We have an object located at a height h

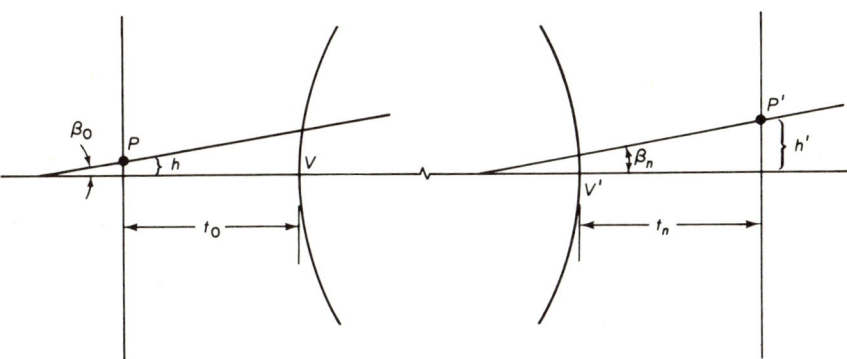

Fig. 3.2. An object-image pair of conjugate points P and P' showing the relevant parameters involved in their specification.

above the optic axis and at a distance t_0 in front of the optical system vertex V. Rays passing through P are transformed by the optical system and form an image at P' a distance t_n from the vertex V' at a height h' above the optic axis. We write the system matrix for our system in the standard form (3.4) and consider the matrix equation for this particular system:

$$\begin{pmatrix} \lambda' \\ h' \end{pmatrix} = \begin{pmatrix} 1 & 0 \\ \mathbf{t}_n & 1 \end{pmatrix} \begin{pmatrix} b & -a \\ -d & c \end{pmatrix} \begin{pmatrix} 1 & 0 \\ \mathbf{t}_0 & 1 \end{pmatrix} \begin{pmatrix} \lambda \\ h \end{pmatrix}. \tag{3.5}$$

We note that t_0 is a negative quantity from the sign convention adopted earlier and we make the substitutions

$$\begin{aligned} \mathbf{t}_0 &= -\mathbf{s}_0 \\ \mathbf{t}_n &= \mathbf{s}, \end{aligned} \tag{3.6}$$

the object of our effort being to find the relationship between s and s_0 in terms of the system parameters a, b, c, and d. Equation (3.5) becomes, on the substitution of (3.6),

$$\begin{pmatrix} \lambda' \\ h' \end{pmatrix} = \begin{pmatrix} 1 & 0 \\ \mathbf{s} & 1 \end{pmatrix} \begin{pmatrix} b & -a \\ -d & c \end{pmatrix} \begin{pmatrix} 1 & 0 \\ -\mathbf{s}_0 & 1 \end{pmatrix} \begin{pmatrix} \lambda \\ h \end{pmatrix}$$

which reduces to

$$\begin{pmatrix} \lambda' \\ h' \end{pmatrix} = \begin{pmatrix} b+a\mathbf{s}_0 & -a \\ \mathbf{s}b+a\mathbf{s}\mathbf{s}_0-d-c\mathbf{s}_0 & c-a\mathbf{s} \end{pmatrix} \begin{pmatrix} \lambda \\ h \end{pmatrix}. \tag{3.7}$$

Equation (3.7) now contains the required relationship between the system parameters. In order to make this more explicit, however, we must consider what is meant by image formation. If P' is in fact the image of P, then all rays passing through P and the system must also pass through P'. This means that we can define an image by the equation

$$h' = \mu h \tag{3.8}$$

where μ is a constant scale factor called the *magnification*. Stated another way, when an image is formed, the position of the image is independent of the angles β_0 of the rays passing through P and, therefore of λ, since all these rays must also pass through P'. In terms of Eq. (3.7) this means that the lower left-hand element in the matrix will be zero, since solution of (3.7) for h' gives

$$h' = (\mathbf{s}b + a\mathbf{s}\mathbf{s}_0 - d - c\mathbf{s}_0)\lambda + (c - a\mathbf{s})h,$$

and since this must be independent of λ:

$$\mathbf{s}b + a\mathbf{s}\mathbf{s}_0 - d - c\mathbf{s}_0 = 0$$

and

$$\mathbf{s} = \frac{d + c\mathbf{s}_0}{b + a\mathbf{s}_0}, \tag{3.9}$$

which gives the desired relationship.

Additionally, the magnification can now be found directly from the relationship between h and h':

$$h' = (c - a\mathbf{s})h = \mu h,$$

so that

$$\mu = c - a\mathbf{s}. \tag{3.10a}$$

We can also exploit the fact that the determinant of the matrix in Eq. (3.7) must be unity. Since the lower left-hand element is zero, the product of the upper left-hand and the lower right-hand elements must be unity. Since the lower right-hand element is the magnification, we can write

$$\mu = \frac{1}{b + a\mathbf{s}_0}. \tag{3.10b}$$

This pair of equations then allows us to determine μ directly either in terms of the image distance alone or in terms of the object distance alone —a very useful pair of alternatives.

The magnification has been defined in Eq. (3.8) as the ratio of h'/h and, if either h or h' is negative (but not both), the magnification will be negative. We can assign a very simple meaning to negative magnification. If the magnification is negative, the object and the image lie on opposite sides of the optic axis. This image inversion is not an uncommon property of optical systems.

IMAGING

The matrix equation between conjugate planes, that is, between pairs of planes that satisfy an object-image relationship, is given by

$$\begin{pmatrix} \lambda' \\ h' \end{pmatrix} = \begin{pmatrix} \dfrac{1}{\mu} & -a \\ 0 & \mu \end{pmatrix} \begin{pmatrix} \lambda \\ h \end{pmatrix}. \tag{3.11}$$

As an illustration of the use of the Eqs. (3.9) and (3.10), we can consider the position and magnification of the image of an object located 100 cm in front of the lens illustrated in Fig. 2.7. We have already found the optical matrix for this lens, and it is given by

$$(s) = \begin{pmatrix} 0.96452 & -0.10805 \\ 0.645 & 0.96452 \end{pmatrix}. \tag{3.12}$$

This gives us the system parameters:

$$a = 0.10805$$
$$b = 0.96452$$
$$c = 0.96452$$
$$d = -0.645$$

by comparison with the standard system matrix (3.4). Note in particular the signs of the various elements.

Since we know the object distance and the refractive index of the medium of the image and object spaces (air $n = 1.000$), we can find the magnification of the image using Eq. (3.10b):

$$\mathbf{s}_0 = s_0 = -100 \text{ cm}$$

$$\mu = \frac{1}{b + a\mathbf{s}_0}$$

$$= \frac{1}{0.96452 + (-100)(0.10805)} = -0.106. \tag{3.13}$$

Thus we see that the image is smaller (0.106 time the object height) and inverted, as we can see from the negative magnification.

Next we can find the image position using Eq. (3.9):

$$\mathbf{s} = s = \frac{d + c\mathbf{s}_0}{b + a\mathbf{s}_0} = \frac{-0.645 + (-100)(0.96452)}{0.6452 + (-100)(0.10805)} = 10.12. \tag{3.14}$$

The image is then formed 10.12 cm to the right of the second vertex of the system.

We can check the internal consistency of our calculation by again calculating the magnification, this time, however, using Eq. (3.10a). We get

$$\mu = c - a\mathbf{s} = 0.96452 - 0.10805(10.12) = -0.108,$$

in good agreement with Eq. (3.13).

In performing the calculation of the image distance as in (3.14), we note that the denominator $(b + a\mathbf{s}_0)$ is just $1/\mu$ so that when performing calculations, if we know μ we can get $\mathbf{s}$ by using

$$\mathbf{s} = \mu(d + c\mathbf{s}_0), \qquad (3.15)$$

and in so doing minimize the arithmetic.

Another example will help with the understanding of the sign convention. We will now take the object to lie 5 cm to the left of the lens which is characterized by the system matrix above. Proceeding as before by using Eq. (3.10b), we find the magnification to be

$$\mu = \frac{1}{b + a\mathbf{s}_0} = \frac{1}{0.9645 + (-5)0.10805} = 2.4,$$

and in this case we find an erect enlarged image. To locate the image we use Eqs. (3.9) or (3.15). Using the latter here, we find

$$\mathbf{s} = s = \mu(d + c\mathbf{s}_0) = 2.4[-0.645 + 0.9645(-5)] = -13.1 \text{ cm}.$$

Here we have a negative image distance, and we will find the image 13.1 cm *to the left* of the second vertex of our system. This means, of course, that the image is virtual.

We have thus seen that the system matrix allows us to relate the image and object position and magnification. In the succeeding chapters we will find that this matrix contains a number of other useful pieces of information about the properties of an optical system.

Problems

3.1 An equiconvex lens in air has an index of 1.6, radii of 5 cm, and a thickness of 2 cm. An object is located on the optic axis 10 cm from the lens. Find the position of the image and the magnification.

3.2 The lens in Problem 3.1 is now placed in water and the object is still 10 cm from the lens on the optic axis. Find the new image position and magnification.

3.3 The lens in Problem 3.1 is inserted in the bottom of a boat to facilitate observation of the bottom. A fish swims by 1 m below the lens. Where does the fish appear to be and what is its apparent size?

3.4 Repeat the above problems with an equiconcave lens with the same parameters.

3.5 Discuss observations made from a boat with a planar glass bottom of thickness t.

Chapter 4

Additional Properties of the System Matrix

In order to investigate the imaging properties of optical systems, we now introduce the concept of the cardinal points or Gaussian points of the optical system. These points occur in pairs and are known as the unit points, the nodal points, and the principal foci of the optical system. The introduction of these cardinal points will allow us to interpret physically the elements a, b, c, and d of the system matrix. The matrix elements of the system matrix will hereafter be referred to as the Gaussian constants of the optical system.

Unit Points and Unit Planes

The *unit planes** are the pair of conjugate planes, one in object space and one in image space, between which the magnification is unity. The *unit points* are the points at which the optic axis and the unit planes intersect. We will denote the unit planes by s_{u0} and s_u, where s_{u0} is the object

---

* The term *principal planes* is also used.

unit plane and s_u the image unit plane. Fortunately, our formulation gives us the relationship between s_{u0} and s_u directly and the magnification through Eqs. (3.10a) and (3.10b):

$$\mu = 1 = c - a\mathbf{s}_u . \tag{4.1a}$$

Similarly,

$$\mu = 1 = \frac{1}{b + a\mathbf{s}_{u0}}, \tag{4.1b}$$

and we have

$$\mathbf{s}_u = \frac{c-1}{a} \tag{4.2a}$$

$$\mathbf{s}_{u0} = \frac{1-b}{a} . \tag{4.2b}$$

Thus the unit planes can be completely specified in terms of the Gaussian constants (system matrix elements).

We now refer to Fig. 4.1. A and A' represent two conjugate planes for our optical system. If we choose to measure distances from the unit planes rather than from the system vertices, we have

$$\begin{aligned} \mathbf{s}_0 &= \mathbf{s}_{u0} + \mathbf{p} \\ \mathbf{s} &= \mathbf{s}_u + \mathbf{q} . \end{aligned} \tag{4.3}$$

In terms of the magnification

$$\mu = c - a\mathbf{s} = c - a\mathbf{s}_u - a\mathbf{q} = 1 - a\mathbf{q} \tag{4.4}$$

and

$$\frac{1}{\mu} = b + a\mathbf{s}_0 = b + a\mathbf{s}_{u0} + a\mathbf{p} = 1 + a\mathbf{p} , \tag{4.5}$$

where we have used Eqs. (4.1a) and (4.1b) to simplify Eqs. (4.4) and (4.5).

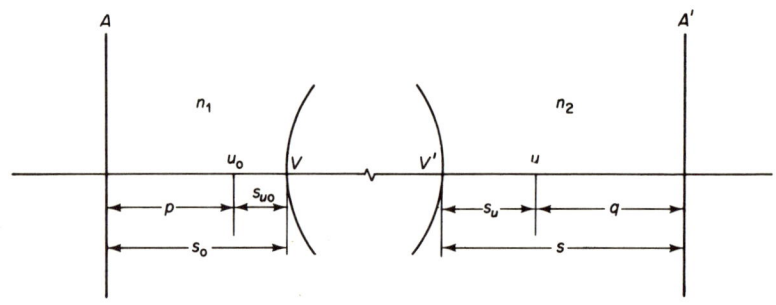

Fig. 4.1. The labeling of conjugate points relative to the unit planes.

If we now substitute in the system matrix written in the form (3.11) we get

$$\begin{pmatrix} 1+a\mathbf{p} & -a \\ 0 & 1-a\mathbf{q} \end{pmatrix} \quad (4.6)$$

with a determinant equal to one. Thus

$$(1+a\mathbf{p})(1-a\mathbf{q}) = 1, \quad (4.7)$$

which can be rearranged to give

$$\frac{1}{\mathbf{q}} - \frac{1}{\mathbf{p}} = a. \quad (4.8)$$

If we now replace the reduced optical distances $\mathbf{p}$ and $\mathbf{q}$ by their equivalent and choose to measure $\mathbf{p}$ as positive from right-to-left, we get, remembering that $\mathbf{p} = P/n_1$ and $\mathbf{q} = q/n_2$:

$$\frac{n_1}{p} + \frac{n_2}{q} = a, \quad (4.9)$$

which can be compared with the lens equation derived in Chapter 1.

Focal Points and Focal Planes

If we now consider a ray entering our optical system from some object at infinity, then all rays from this object enter the system parallel to the optic axis. The image point and its associated plane determine the image-side principal focus as illustrated in Fig. 4.2. Operating with Eq. (4.7) in the form

$$(1-a\mathbf{q}) = \frac{1}{1+a\mathbf{p}}, \quad (4.10)$$

and letting $p \to \infty$, we get

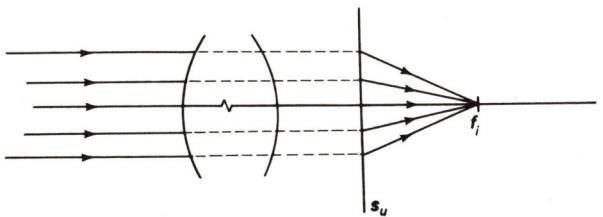

Fig. 4.2. The image-side principal focus.

$$\mathbf{q} = \mathbf{f}_i = \frac{1}{a}, \tag{4.11}$$

which allows us to interpret the Gaussian constant a directly. This constant is in fact the reciprocal of the image-side focal length measured from the unit plane $\mathbf{s}_u$.

We can define the object-side principal focus $\mathbf{f}_0$ as the object point which gives an image at infinity, i.e., that point for which all rays passing through the optical system emerge parallel to the optic axis as illustrated in Fig. 4.3. Following the same procedure as before beginning with Eq. (4.7), we get

$$1 + a\mathbf{p} = \frac{1}{(1 - a\mathbf{q})}, \tag{4.12}$$

and letting $\mathbf{q} \to \infty$, we find

$$\mathbf{f}_0 = \mathbf{p} = -\frac{1}{a}. \tag{4.13}$$

This means that so long as a is positive, the object-side principal focus will lie to the left of the object-side principal plane. Again we see that the Gaussian constant a has a simple meaning.

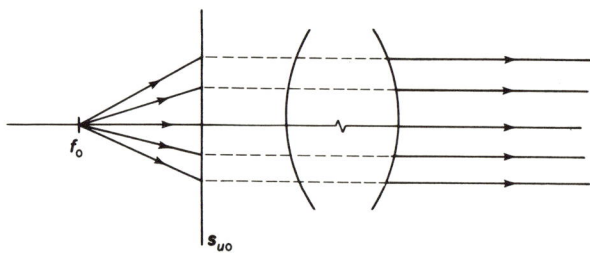

Fig. 4.3. The object-side principal focus.

Examination of Eqs. (4.11) and (4.13) shows that the principal foci are each found at the same distance from the respective principal plane and symmetrically placed with regard to this principal plane so long as the external medium surrounding the optical system is uniform.

We can now determine the position of the unit planes and the principal foci for the example illustrated in Fig. 2.7 and for which the system matrix is given by

$$(s) = \begin{pmatrix} 0.96452 & -0.10805 \\ 0.645 & 0.96452 \end{pmatrix}. \tag{4.14}$$

Using Eqs. (4.2), we get

$$\mathbf{s}_u = \frac{c-1}{a} = \frac{0.96452-1}{0.10805} = -0.33$$

$$\mathbf{s}_{u0} = \frac{1-b}{a} = \frac{1-0.96452}{0.1} = +0.33,$$

and in this case the unit planes are effectively at the system vertices since, even taking account of the index for some external medium for $\mathbf{s}_u$ and $\mathbf{s}_{u0}$ to convert them into distances from reduced distances, the unit planes will lie within a third of a centimeter of the vertices. The positions are illustrated in Fig. 4.4 to indicate their placement.

The focal planes are found using Eqs. (4.11) and (4.13), and we get

$$\mathbf{f}_i = \frac{1}{a} = \frac{1}{0.10802} = 9.26$$

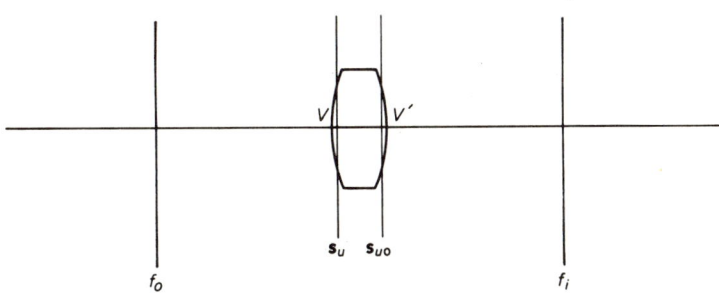

Fig. 4.4. The location of the unit planes and the foci for a thick, double-convex lens.

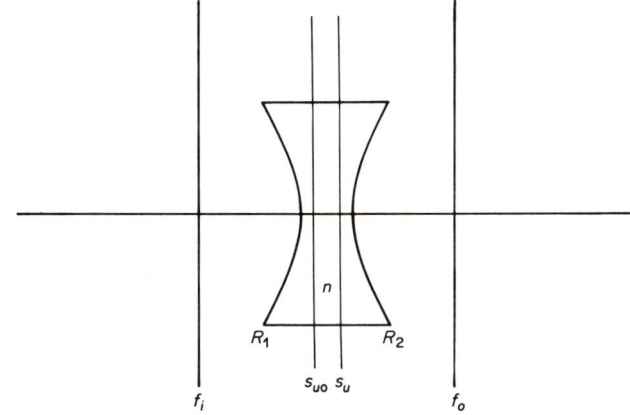

Fig. 4.5. The location of the unit planes and the foci for a thick, double-concave lens.

CHAPTER 4 ADDITIONAL PROPERTIES OF THE SYSTEM MATRIX

and

$$\mathbf{f}_0 = -\frac{1}{a} = -\frac{1}{0.10802} = -9.26.$$

Both $\mathbf{f}_0$ and $\mathbf{f}_i$ (like $\mathbf{s}_{u0}$ and $\mathbf{s}_u$) are reduced distances, so that before we can actually draw their positions we need to know the index of the medium in which the lens is immersed. However, if we simply assume that the medium is air, we can give their positions as is done in Fig. 4.4.

Another example will help to illustrate the symmetry of placement of the unit planes and foci. Figure 4.5 show the lens chosen for this new example, and Table 4.1 gives the necessary data to generate the system matrix.

Table 4.1

	R (cm)	n	n'	t
1	-6.00	1.000	1.500	2.000
2	6.00	1.500	1.000	

This is a double concave lens in an air medium. The system matrix is given by

$$\begin{pmatrix} 1 & -\dfrac{1.000-1.500}{6.00} \\ 0 & 1 \end{pmatrix} \begin{pmatrix} 1 & 0 \\ \dfrac{2.00}{1.500} & 1 \end{pmatrix} \begin{pmatrix} 1 & -\dfrac{1.500-1.000}{-6.00} \\ 0 & 1 \end{pmatrix},$$

which reduces to

$$(s) = \begin{pmatrix} 1.11111 & 0.17593 \\ 1.33334 & 1.11111 \end{pmatrix} \quad (4.15)$$

$$|s| = 1.000.$$

The Gaussian constants are given by

$$a = -0.17593$$
$$b = 1.11111$$
$$c = 1.11111$$
$$d = -1.33334,$$

which can be used to find the unit planes and principal foci. Using Eqs. (4.2), we have

$$\mathbf{s}_u = \frac{c-1}{a} = \frac{1.11111-1.00}{-0.17593} = -0.632$$

$$\mathbf{s}_{u0} = \frac{1-b}{a} = \frac{1.00-1.11111}{-0.17593} = 0.632,$$

and with Eqs. (4.11) and (4.13) we get

$$\mathbf{f}_i = \frac{1}{a} = -5.685$$

and

$$\mathbf{f}_0 = -\frac{1}{a} = 5.685.$$

These quantities are shown in Figure 4.5. This is a negative or diverging lens. In such a case the image-side focal point lies to the left of $\mathbf{s}_{u0}$. This is in contrast to the situation with the converging lens illustrated in Fig. 4.4 where the image-side focus lies to the right of the unit plane and the object-side focus to the left. With a negative lens one finds virtual images are formed for real objects. This will be further illustrated when we consider images in some later examples.

Nodal Points

Perspective is an important feature in projection systems. Ideally, the relative angles of view presented by the image to the eye should be similar to those one finds in viewing the object itself. We then want to know if a pair of conjugate points can be found such that a ray passing through the first point and then through the lens passes through a second point in such a way that it makes the same angle with the optic axis. Conjugate points having this property are called nodal points. We will label these points $\mathbf{s}_N$ and $\mathbf{s}_{N0}$, where the latter is the object-side nodal point. Figure 4.6 illustrates the geometry involved where the angles α and α' are identical.

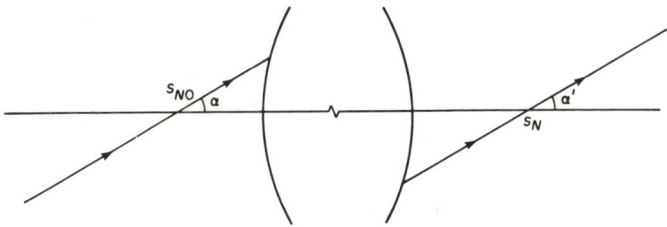

Fig. 4.6. The nodal points.

CHAPTER 4 ADDITIONAL PROPERTIES OF THE SYSTEM MATRIX

The nodal points are easily found in the matrix theory. If we take the matrix equation between conjugate planes (3.11), we find

$$\begin{pmatrix} \lambda' \\ h' \end{pmatrix} = \begin{pmatrix} \dfrac{1}{\mu} & -a \\ 0 & \mu \end{pmatrix} \begin{pmatrix} \lambda \\ h \end{pmatrix}. \tag{4.16}$$

What we want to find is the pair of points for $h=0$ such that $\lambda'=\lambda$. The relationship between λ' and λ when $h=0$ is given by

$$\frac{\lambda'}{\lambda} = \frac{1}{\mu}. \tag{4.17}$$

We recall that λ and λ' are not angles but are reduced angles and*

$$\lambda = n\alpha \tag{4.18a}$$
$$\lambda' = n'\alpha', \tag{4.18b}$$

so that

$$\frac{\alpha'}{\alpha} = \frac{n}{n'\mu}. \tag{4.19}$$

To find the nodal points we take

$$\frac{n}{n'\mu} = 1, \tag{4.20}$$

which gives

$$\mu = \frac{n}{n'} = c - \mathbf{s}_N a \tag{4.21a}$$

and

$$\mu = \frac{n}{n'} = \frac{1}{b + \mathbf{s}_{N0} a}. \tag{4.21b}$$

The linearity of these equations assures us that within the limits of this theory only one such set of points can exist. Solving Eqs. (4.21) then allows us to find $\mathbf{s}_N$ and $\mathbf{s}_{N0}$ as an expression involving the indices of refraction and the Gaussian constants of the system:

$$\mathbf{s}_N = \frac{c - (n/n')}{a} \tag{4.22a}$$

$$\mathbf{s}_{N0} = \frac{(n'/n) - b}{a}. \tag{4.22b}$$

A direct consequence of these equations is that for a lens in a homogeneous medium the nodal points and the unit points coincide.

---

* n is the index in object space and n' is the index in image space.

We can see that in our previous illustrations where the lenses were in air, the nodal points would lie within the lenses.

Graphical Image Construction

We have now found the cardinal points of optical systems in terms of the Gaussian constants of the optical system. The cardinal points, in particular the unit planes and the focal points, can be used to graphically construct images for a given object.

To find the image associated with some object, two rays are generally traced. Figure 4.7 is a simplified diagram showing just the focal points and the unit planes for a converging lens such as is given in Fig. 4.4. The first ray chosen is the one parallel to the optic axis from A_0'. This ray strikes s_{u0} at point B_1'. The definition of the unit planes tells us that B_1 is conjugate to a point B_1' which has the same position on s_u relative to the optic axis (unit magnification). In addition, a ray entering a system parallel to the optic axis must pass out through f_i so that we can completely specify the path of this ray by knowing the position of the unit planes and the focal points.

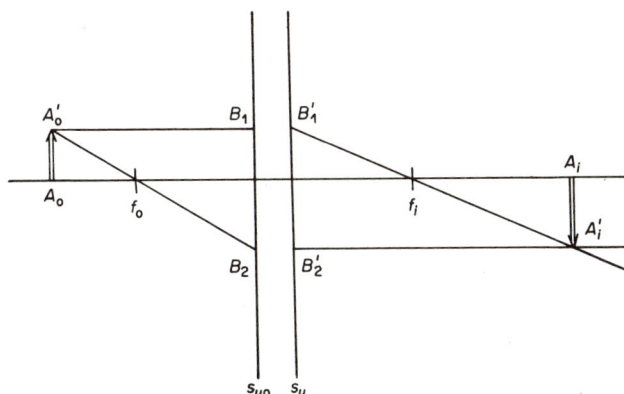

Fig. 4.7. Graphical image construction for a positive lens.

As our second ray we choose the ray from A_0' that passes through f_0 and intersects s_{u0} at B_2. This ray will be conjugate to a point B_2' identically situated on s_u. We can, by using the property that a ray through f_0 must leave the optical system parallel to the optic axis, specify the final path of this ray. The image point A_i' conjugate to A_0' will lie at the intersection of the two rays. The point conjugate to A_0 will lie on the axis directly below (above) $A_i{}'$, and we have found the image of the object A.

Note that if the vertical scale in Fig. 4.7 is greatly enlarged, such a large object so near to the lens would generally not fall within the limits of the paraxial-ray assumption.

We can verify our construction by calculation. For Fig. 4.4, f_0 lies 9.26 cm to the left of the vertex, so we will take A to be 12 cm from the vertex. The image point can be found by using Eq. (3.9):

$$s = \frac{d+c\mathbf{s}_0}{b+a\mathbf{s}_0} \approx 36 \text{ cm},$$

where the external medium is air and the Gaussian constants are found from the matrix (3.13). The result is consistent with the diagram which is not drawn precisely to scale. The magnification should be negative since the image is inverted, and by using Eq. (3.10a) we find

$$\mu = c - a\mathbf{s} = 0.96452 + (-36)(0.10805) \approx -3,$$

so that the image is inverted and larger. Of course, the image is real since it lies to the right of the optical system and since, as we can see from Fig. 4.7, the rays actually converge at the image point.

There are a number of different situations, such as virtual objects and virtual images, that may arise, but all can be treated in the same way as in the previous example. To illustrate this we examine Fig. 4.8 which

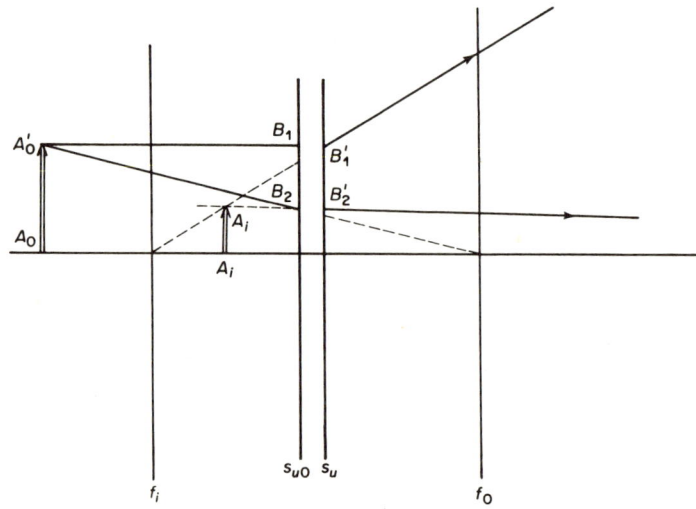

Fig. 4.8. Graphical image construction for a negative lens.

gives the unit planes and principal foci for the diverging lens treated in the last section (Fig. 4.5).

The ray from A_0' parallel to the optic axis strikes s_{u0} at B_1 and appears at the same position on s_u after passing through the system. This ray must pass through f_i, and since in this case f_i lies to the left of s_u, it is the backward projection that determines the direction of the ray. The ray itself does not pass through f_i. The ray from A_0' which is directed toward f_0 strikes s_{u0} at B_2 and then appears at B_2'. It then passes out of the system parallel to the optic axis. The backward projection of this ray and its intersection with the first ray projection determine the image position. The image in this case is erect, virtual, and smaller.

Care must be taken in the construction of ray diagrams to ensure that the image and object unit planes and foci are clearly marked. It sometimes occurs that s_u lies to the left of s_{u0}, and in such cases one must be extremely careful to follow the principles we have set down for constructing ray diagrams explicitly. In the discussion of optical instruments later in this book such a case will arise for the Ramsden eyepiece.

The Thin Lens

A special case of the system matrix arises when we treat the thin lens. In Chapter 1 (Fig. 1.9) this case was discussed using the algebraic theory set down in that chapter. We consider now the lens illustrated in Fig. 4.9. We will specify the power of surface 1 by a_1 and that of surface 2 by a_2. The system matrix for this simple lens is given by

$$(S) = \begin{pmatrix} 1 & -a_2 \\ 0 & 1 \end{pmatrix} \begin{pmatrix} 1 & 0 \\ \dfrac{t}{n} & 1 \end{pmatrix} \begin{pmatrix} 1 & -a_1 \\ 0 & 1 \end{pmatrix}. \tag{4.23}$$

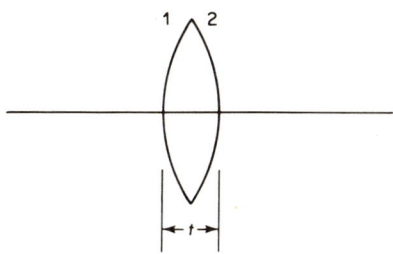

Fig. 4.9. The thin lens. A lens is called thin when t becomes vanishingly small or is small relative to the other parameters of the problem.

CHAPTER 4 ADDITIONAL PROPERTIES OF THE SYSTEM MATRIX

For thin lenses we allow $t \to 0$, i.e., we use precisely the same assumption as in Chapter 1 when we let $\delta \to 0$ for the two surfaces forming the lens. In this case the translation matrix in (4.23) become the unit matrix, the identity element in the algebra of these matrices, and (S) is now given by

$$(S) = \begin{pmatrix} 1 & -a_2 \\ 0 & 1 \end{pmatrix}\begin{pmatrix} 1 & -a_1 \\ 0 & 1 \end{pmatrix}, \tag{4.24}$$

which reduces to

$$(S) = \begin{pmatrix} 1 & -(a_1+a_2) \\ 0 & 1 \end{pmatrix}. \tag{4.25}$$

Thus we see that the system matrix has a particularly simple form for a thin lens.

If we examine Eq. (4.25) we find, first of all, that the unit planes gotten from the solutions of Eqs. (4.2),

$$\begin{aligned} \mathbf{s}_{u0} &= \frac{c-1}{a} = \frac{1-1}{a_1+a_2} = 0 \\ \mathbf{s}_u &= \frac{1-b}{a} = \frac{1-1}{a_1+a_2} = 0, \end{aligned} \tag{4.26}$$

lie in coincidence with the lens (considered as a plane). Thus the focal length of the lens will be measured from the lens itself (Eq. 4.9). Combining (4.9) with the physical interpretation of a as $1/f$, we get

$$\frac{n_1}{p} + \frac{n_2}{q} = a = \frac{1}{f}. \tag{4.27}$$

If we apply Eq. (4.9) to a thin lens, $a = a_1 + a_2$, and if a_1 and a_2 are given by

$$a_1 = \frac{n-n_1}{R_1}$$

$$a_2 = \frac{n_2-n}{R_2},$$

we can write

$$\frac{1}{f} = a_1 + a_2 = \frac{n-n_1}{R_1} + \frac{n_2-n}{R_2}. \tag{4.28}$$

For a lens in air, which was the case treated in Chapter 1, $n_1 = n_2 = 1$, and

$$\frac{1}{f} = (n-1)\left(\frac{1}{R_1} - \frac{1}{R_2}\right). \tag{4.28a}$$

This is the lensmaker's equation (1.21), and we have again demonstrated the consistency between the matrix formulation and the algebraic treatment developed in Chapter 1. Thus, given the focal length for a thin lens, one can write down the system matrix directly, and this turns out to be worthwhile since in many thin lens problems we are given the focal length of the lens rather than the lens parameters.

The focal length of two thin lenses in contact, i.e., with zero spacing between them, can be gotten directly from their system matrix:

$$(S) = \begin{pmatrix} 1 & -\frac{1}{f_2} \\ 0 & 1 \end{pmatrix} \begin{pmatrix} 1 & 0 \\ 0 & 1 \end{pmatrix} \begin{pmatrix} 1 & -\frac{1}{f_1} \\ 0 & 1 \end{pmatrix}$$

$$= \begin{pmatrix} 1 & -\left(\frac{1}{f_1}+\frac{1}{f_2}\right) \\ 0 & 1 \end{pmatrix},$$

so that for two thin lenses in contact:

$$\frac{1}{f} = \frac{1}{f_1} + \frac{1}{f_2}. \tag{4.29}$$

This is a standard result found in all texts treating thin lenses and illustrates the facility with which we can generate results for lens combinations using the matrix formalism.

Changing the External Medium

System matrices are generally developed for specific problems, and these matrices depend on the medium *surrounding* the lens. If we write out the product for some system (Fig. 4.10), we get

$$(S) = \begin{pmatrix} 1 & -\frac{n_i - n_{j-1}}{R_j} \\ 0 & 1 \end{pmatrix} \cdots \begin{pmatrix} 1 & -\frac{n_1 - n_0}{R_1} \\ 0 & 1 \end{pmatrix}. \tag{4.30}$$

Suppose that we want to change the medium at the object side from n_0 to air $n=1.00$. A simple procedure takes the following form: Figure 4.11 shows the entrance side of the system with surface 1, we imagine a thin lens of index $n=n_0$ with both radii of curvature equal to R_1 to be at surface 1 and then let the thickness of this lens go to zero. Taking only the final matrix of (4.30), we extend expression (4.30) to include this thin lens:

50 CHAPTER 4 ADDITIONAL PROPERTIES OF THE SYSTEM MATRIX

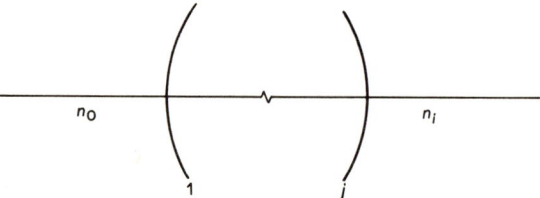

Fig. 4.10. A general optical system separating two media with different optical properties.

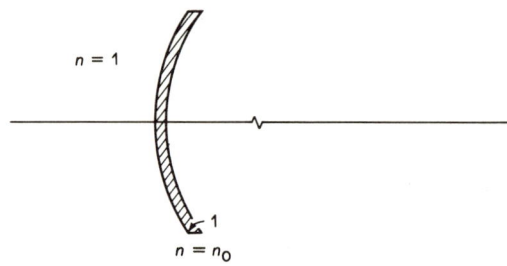

Fig. 4.11. The thin lens used in the conversion of optical matrices in situations where the index of the external medium is allowed to change.

$$\cdots \begin{pmatrix} 1 & -\dfrac{n_1-n_0}{R_1} \\ 0 & 1 \end{pmatrix} \begin{pmatrix} 1 & 0 \\ \dfrac{t}{n_0} & 1 \end{pmatrix} \begin{pmatrix} 1 & -\dfrac{n_0-1}{R_1} \\ 0 & 1 \end{pmatrix}. \quad (4.31)$$

We now let $t \to 0$. The central matrix is an identity element and therefore can be dropped; we get as a product

$$\cdots \begin{pmatrix} 1 & -\dfrac{n_1-1}{R_1} \\ 0 & 1 \end{pmatrix}, \quad (4.32)$$

which is in the form required of the first refraction matrix for a system with air at the first surface. A similar argument can be used on the final surface. Of course, as long as we do not attempt to rearrange the matrices, the order in which multiplication is carried out is immaterial, so that give a system matrix derived for a system in a single medium of index n_0, if we wish to find the matrix derived for a system in a medium n_k, we can postmultiply by (4.32) and premultiply by the similar matrix for the jth face of the system, and we get

$$(S)_{n_k n_k} = \begin{pmatrix} 1 & -\dfrac{n_k - n_0}{R_j} \\ 0 & 1 \end{pmatrix} (S)_{n_0 n_0} \begin{pmatrix} 1 & -\dfrac{n_0 - n_k}{R_1} \\ 0 & 1 \end{pmatrix}, \tag{4.33}$$

where the subscripts on the system matrices are used to distinguish the external medium. For systems other than those in air it is recommended that suffixes be written to identify the external medium.

We have now examined the elements of the system matrix in some detail. We now must deal with the questions of application of this formalism to practical optical systems.

Problems

4.1 A plano-convex lens 3.5 cm thick is made of glass with index 1.60. If the spherical surface has a radius of 5 cm, find the unit planes and the focal points. What is the effect of reversing the lens on the positions of these points? If the lens is allowed to become thin, what happens to the unit planes and the focal points in both orientations?

4.2 A clear ballon filled with air is used as an underwater lens. Discuss the imaging properties and the positions of the various cardinal points.

4.3 Where are the nodal points for the lens in Problem 4.1.

4.4 Two thin lenses with focal lengths $+10$ and -10 cm are located 10 cm apart in air. Find the cardinal points for both possible combinations. The space between the lenses is now filled with water. What is the effect on the cardinal points?

4.5 Find the cardinal points for the lens in problem 2.7. Construct the image graphically to verify the results.

4.6 A symmetric convex lens with radii equal to 10 cm is 3 cm thick and is made of glass $n=1.6$. If the lens is in air, find the cardinal points. If the lens is built into the wall of a tank containing a fluid $n=1.4$, find the new cardinal points.

4.7 Two thin lenses have the following parameters: L_1, $r_1 = 16$ cm,

$r_2 = 24$ cm, $n = 1.6$; L_2, $r_1 = 32$ cm, $r_2 = 48$ cm, $n = 1.5$. Find the focal length of the combination of these two lenses in contact and write the system matrix for the combination. What is the effect on the properties of the system of reversing the order of the lenses?

4.8 Two thin lenses with focal lengths of 5 cm and 10 cm are 5 cm apart in air. Find the cardinal points and the image position and magnification for an object 1 cm high located on the optic axis 20 cm to the left of the system. Verify the result by graphical image construction.

4.9 What is the minimum amount of information about an unknown complex lens system that will allow you to write the system matrix?

Chapter 5

Mirrors

In Chapter 1 the spherical mirror was treated as a case different from that of a spherical refracting surface. The purpose of this chapter is to see how spherical reflecting surfaces can be treated in the matrix formalism. Figure 5.1 shows an image-object construction for a concave spherical reflecting surface. The equation governing the object and image relationships was found in Chapter 1 to be

$$\frac{1}{p}+\frac{1}{q}=\frac{1}{f},\qquad(5.1)$$

where

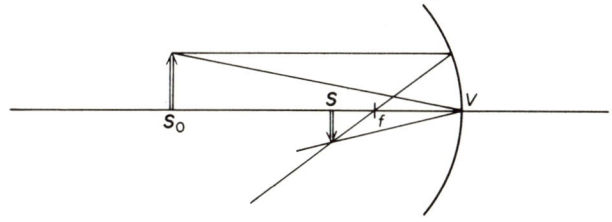

Fig. 5.1. Geometrical construction of the image for a concave spherical mirror.

$$\frac{1}{f} = \frac{2}{R} \tag{5.2}$$

and p is positive for objects to the left of the mirror and q is positive for images to the left of the mirror. For concave mirrors R was always taken as positive. The image construction is straightforward. In Fig. 5.1 we use the ray to the vertex V, which must make equal angles with the optic axis, and the ray parallel to the optic axis, which must pass out through the focal point.

We next set up the system matrix for this situation. This is a single surface system, so that we get

$$(S) = \begin{pmatrix} 1 & -\dfrac{n'-n}{R} \\ 0 & 1 \end{pmatrix}, \tag{5.3}$$

and for the moment we choose to ignore the fact that the reflected ray passes through a medium with the same index as the incident ray. Actually, in all previous examples we have only considered rays traveling from left to right, and in this case, the reflected ray travels in the opposite sense. Additionally, in terms of the formalism we have thus far developed, taking $n=n'$ would give a unit matrix which would give an improper result and the only free variable with which we can deal within the framework of the theory is n'.

We now take the system matrix (5.3) with Gaussian elements:

$$a = \frac{n'-n}{R}$$
$$b = 1$$
$$c = 1$$
$$d = 0$$

and substitute into Eq. (3.9), which gives the object-image relationship for a system matrix:

$$\mathbf{s} = \frac{\mathbf{s}_0}{1 + \dfrac{n'-n}{R}\mathbf{s}_0}, \tag{5.4}$$

which can be rearranged to give

$$\frac{1}{\mathbf{s}} - \frac{1}{\mathbf{s}_0} = \frac{n'-n}{R}. \tag{5.5}$$

MIRRORS

Equation (5.5) appears to be quite similar to Eq. (5.1). If we change the reduced distances $\mathbf{s}_0$ and $\mathbf{s}$ into unweighted distances, we get

$$-\frac{n}{\mathbf{s}_0}+\frac{n'}{\mathbf{s}}=\frac{n'-n}{R}. \tag{5.6}$$

Equation (5.1) is independent of the medium. In the derivation of the mirror equation only the equal angle theorem is used and this is not altered when the mirror is in a medium with an index different from one. We can therefore state that $|n'|=|n|$, and since we know that the left-hand side of (5.6) must be equal to $2/R$, we take $n'=-n$. Equation (5.6) then takes the form

$$\frac{1}{s_0}+\frac{1}{s}=\frac{2}{R}. \tag{5.7}$$

In order to assure ourselves that Eq. (5.7) is valid within our previously established sign convention, we can take the example illustrated in Fig. 5.1, where $s_0=-100$ cm and $R=-50$ cm. Solving Eq. (5.7) by using these values, we find

$$-\frac{1}{100}+\frac{1}{s}=\frac{2}{-50},$$

from which we get

$$s=-\frac{100}{3}\text{ cm},$$

a result consistent with Fig. 5.1, if we remember that distances measured to the left of the vertex are negative.

We also see that the image is inverted and smaller, so that the magnification μ must be negative and less than unity. Using Eq. (3.10a) we get

$$\mu = c - a\mathbf{s} = 1 - \left(\frac{-2n}{-50}\right)\left(\frac{100}{3}\Big/(-n)\right) = -\frac{1}{3},$$

and with Eq. (3.10b) we get

$$\mu = \frac{1}{b+a\mathbf{s}_0} = \frac{1}{1+\left(\frac{-2n}{-50}\right)\left(\frac{-100}{n}\right)} = -\frac{1}{3},$$

again consistent with the Fig. 5.1.

The Plane Mirror

In the case of a plane mirror $R \to \infty$ and the system matrix goes to

$$(S) = \begin{pmatrix} 1 & -\dfrac{n'-n}{R} \\ 0 & 1 \end{pmatrix} \to \begin{pmatrix} 1 & 0 \\ 0 & 1 \end{pmatrix}. \tag{5.8}$$

The plane mirror system matrix is just the unit matrix. From Eqs. (5.8) and (3.9) we get the object-image relation for the plane mirror in the form

$$\mathbf{s} = \frac{0+\mathbf{s}_0}{1+0\mathbf{s}_0} = \mathbf{s}_0, \tag{5.9}$$

thus $\mathbf{s}=\mathbf{s}_0$ is the required equation. When these are converted from reduced distances into direct lengths we find

$$\frac{s_0}{n} = \frac{s}{-n}, \tag{5.10}$$

and the relationship between the direct distances becomes $s=-s_0$. The image lies behind the mirror and is virtual since for real objects $s_0<0$ and therefore $s>0$. Image and object are symmetrically placed about the mirror (Fig. 5.2).

The magnification is

$$\mu = c - a\mathbf{s} = \frac{1}{b+a\mathbf{s}_0} = 1,$$

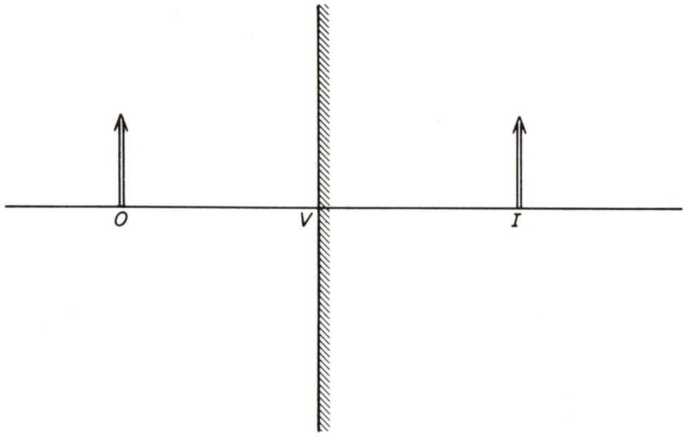

Fig. 5.2. The object and image for a plane mirror.

so that we have an erect object with unit magnification for any position of the real object.

The Convex Mirror—An Example

We will now treat a convex mirror with a radius of curvature $R = 50$ cm and $s_0 = -100$ cm as illustrated in Fig. 5.3. The same principles are used in the geometrical construction of the image as were used in Fig. 5.1. We use two principal rays: that parallel to the optic axis, which

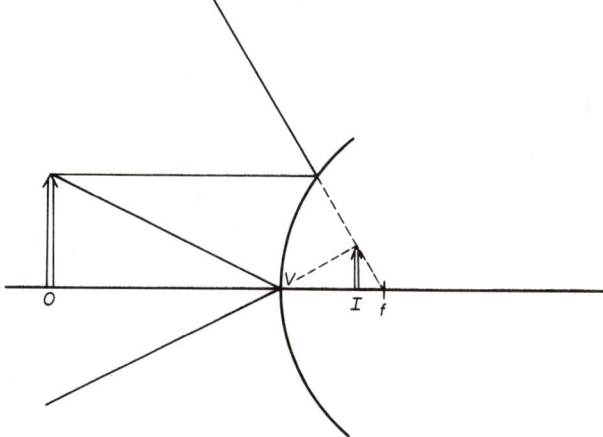

Fig. 5.3. Graphical construction of the object for a convex (negative) spherical mirror.

must appear to pass through the focus, and that which strikes the mirror at the vertex and therefore makes equal angles with the optic axis at that point. From the construction we expect to find an erect, smaller, virtual image located between the focus and the mirror.

For this system we have as system matrix

$$(S) = \begin{pmatrix} 1 & -\dfrac{-2n}{R} \\ 0 & 1 \end{pmatrix} = \begin{pmatrix} 1 & -\dfrac{-2n}{50} \\ 0 & 1 \end{pmatrix}, \quad (5.11)$$

and the Gaussian constants

$$a = \frac{-2n}{50}$$

$$b = 1$$
$$c = 1$$
$$d = 0.$$

To find the image position we use Eq. (3.9) and find

$$\mathbf{s} = \frac{s}{-n} = \frac{d + c\mathbf{s}_0}{b + a\mathbf{s}_0} = \frac{0 + \dfrac{-100}{n}}{1 + \dfrac{-2n}{50}\dfrac{-100}{n}}$$

and

$$s = 20 \text{ cm}.$$

Since $f = R/2 = 25$ cm we have, as expected, found a virtual image in the anticipated position.

The magnification is given by Eq. (3.10a):

$$\mu = c - a\mathbf{s} = 1 - \left(\frac{-2n}{50}\right)\left(\frac{20}{-n}\right) = 0.2,$$

and we find a positive magnification (image erect) and a smaller image.

A Lens with a Reflecting Surface

As a final example of the use of matrices for reflecting surfaces, we treat the case of a partially reflecting lens. Figure 5.4 illustrates the physical system. We have a spherical glass lens of radius R and index n_0 in an air medium. The object lies to the left of the system and the incoming ray is refracted by the surface at vertex V_1, translated to the vertex V_2 where it is reflected, again translated back to the surface V_3, and again refracted. The system matrix is made up of two refraction matrices, two translation

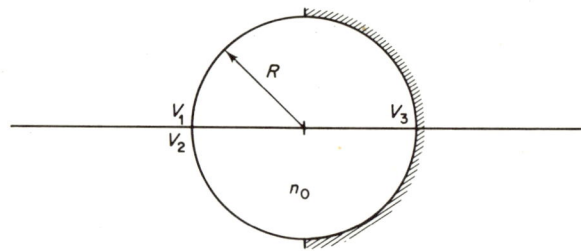

Fig. 5.4. A lens with one reflecting surface.

matrices, and one reflection matrix so that we write:

$$(S) = \mathscr{R}_3 \mathscr{T}_2 \mathscr{R}_2 \mathscr{T}_1 \mathscr{R}_1, \qquad (5.12)$$

and we will construct each of these matrices separately so as to comment on the signs of the various elements.

$\mathscr{R}_1$ is a straightforward refraction matrix between air and glass, $n = n_0$, where the radius of curvature of the sphere is positive:

$$\mathscr{R}_1 = \begin{pmatrix} 1 & -\dfrac{n_0-1}{R} \\ 0 & 1 \end{pmatrix}. \qquad (5.13)$$

$\mathscr{T}_1$ is a translation matrix through the sphere, a distance of $2R$:

$$\mathscr{T}_1 = \begin{pmatrix} 1 & 0 \\ \dfrac{2R}{n_0} & 1 \end{pmatrix}. \qquad (5.14)$$

$\mathscr{R}_2$ is the reflection matrix at a surface with a negative radius of curvature:

$$\mathscr{R}_2 = \begin{pmatrix} 1 & -\dfrac{(-n_0)-n_0}{-R} \\ 0 & 1 \end{pmatrix}. \qquad (5.15)$$

$\mathscr{T}_2$ is a translation back to the front surface, and

$$\mathscr{T}_2 = \begin{pmatrix} 1 & 0 \\ \dfrac{-2R}{-n_0} & 1 \end{pmatrix}. \qquad (5.16)$$

We note that the distance measured, V_2 to V_3, is negative and the index is still negative since the ray is traveling from right to left. Finally, R_3 is given by

$$\mathscr{R}_3 = \begin{pmatrix} 1 & -\dfrac{(-1)-(-n_0)}{R} \\ 0 & 1 \end{pmatrix}, \qquad (5.17)$$

where both refractive indices are negative because of the reversed direction of the ray, but R remains positive.

Combining Eqs. (5.13) through (5.17) as required in Eq. (5.12), we get

$$(S) = \begin{pmatrix} 1 & -\dfrac{(-1)-(-n_0)}{R} \\ 0 & 1 \end{pmatrix} \begin{pmatrix} 1 & 0 \\ \dfrac{-2R}{-n_0} & 1 \end{pmatrix} \begin{pmatrix} 1 & -\dfrac{(-n_0)-n_0}{-R} \\ 0 & 1 \end{pmatrix} \cdots$$

$$\times \begin{pmatrix} 1 & 0 \\ \dfrac{2R}{n_0} & 1 \end{pmatrix} \begin{pmatrix} 1 & -\dfrac{n_0-1}{1} \\ 0 & 1 \end{pmatrix}, \qquad (5.18)$$

which gives

$$(S) = \begin{pmatrix} \dfrac{n_0-4}{n_0} & \dfrac{2n_0-4}{Rn_0} \\ -\dfrac{4R}{n_0} & \dfrac{n_0-4}{n_0} \end{pmatrix} \qquad (5.19)$$

$$|S| = 1,$$

and this matrix, with the properly assigned values for R and n_0, will then give the optical properties of the system pictured in Fig. 5.4.

For example, if we take $n_0 = 1.5000$ and $R = 5$ cm, an object at a distance 50 cm from V_1 will form an image, and the position of the image can be found as usual by using Eq. (3.9). The Gaussian constants of this system are

$$a = \frac{4-2n_0}{Rn_0} = 1.3334$$

$$b = \frac{n_0-4}{n_0} = 1.6667$$

$$c = \frac{n_0-4}{n_0} = 1.6667$$

$$d = \frac{4R}{n_0} = 13.3334.$$

We note that $\mathbf{s}_0 = s_0$ and $\mathbf{s} = s$ since the external medium is air, and substituting in Eq. (3.9) we find

$$s = \frac{13.3334 + 1.6667(-50)}{1.6667 + 1.3334(-50)}$$

$$= 1.08 \text{ cm}.$$

s is positive and its position is measured relative to V_3, so it lies 1.08 cm inside the front surface of the sphere. The magnification given by Eq.

(3.10a) yields

$$\mu = c - a\mathbf{s} = 1.6667 - (1.3334)(1.08)$$
$$= 0.2266,$$

so the image is erect and smaller.

Problems

5.1 The radius of curvature of a concave spherical mirror is 50 cm. An object 3 cm high is located in front of the mirror at a distance of (a) 60 cm, (b) 30 cm, (c) 15 cm, and (d) 5 cm. Find the position and the size of each of the images. Repeat the problem for a convex mirror with the same radius.

5.2 Verify one of the results of Problem 5.1 by graphical construction for the concave and convex mirror cases.

5.3 A concave spherical mirror is used to focus an object on a screen located 120 cm from the object. A magnification of -16 is desired. What should be the radius of the mirror and where should the mirror be located?

5.4 A thin lens of index 1.50 is symmetric and positive and has radii of curvature of 12 cm. One side is silvered. Find the system matrix for light entering through the unsilvered side.

5.5 A fish swims in a glass fish bowl 1 ft in diameter. If a plane mirror is located 1 ft behind the fish bowl, where does the image of the fish appear?

5.6 A thin lens with focal length -14.5 cm is placed 3 cm from a convex spherical mirror with radius 15 cm. Find the system matrix.

5.7 The curved surface of a plano-convex lens is silvered. Discuss the optical properties of this lens.

Chapter 6

Stops and Pupils; Chromatic Aberration

Before proceeding to study a number of specific optical instruments, we are going to consider two factors affecting the quality of optical images, both of which can be handled within the paraxial-ray assumption. Stops and pupils are related to the light-gathering properties of optical systems. We will be interested in these discussions in just how much light from a given source will pass through a given optical system, and we will examine the factors that limit the light. We will also examine the effect of dispersion, the wavelength dependence of the refractive index of the optical system on the formation of images.

Stops and Pupils

Two very important questions in the study of optical systems concern the brightness of the image and the size of the objects which can be imaged by the optical system, i.e., the field of view. Both of these questions require us to examine that portion of the optical system that limits the light passing through the system.

64   CHAPTER 6   STOPS AND PUPILS; CHROMATIC ABERRATION

Figure 6.1 illustrates an *aperture stop* AS for a single thin lens. We have illustrated the stop as a hole in an opaque screen. Only those rays that pass through the central part of the lens help to form an image on the focal plane. The peripheral rays that strike near the lens edge cannot contribute to the image since they are blocked by the aperture stop.

Figure 6.2 illustrates a *field stop*, FS. The effect of this stop is to limit the size of the object that can be imaged on the focal plane. Rays (solid lines) from objects lying far off the optic axis are cut off before the focal plane FP by the stop. Only those rays (dotted lines) arising from objects relatively close to the optic axis will be imaged.

These stops need not necessarily be explicitly inserted in the system as illustrated in the figures. They may, and often do, arise because of the

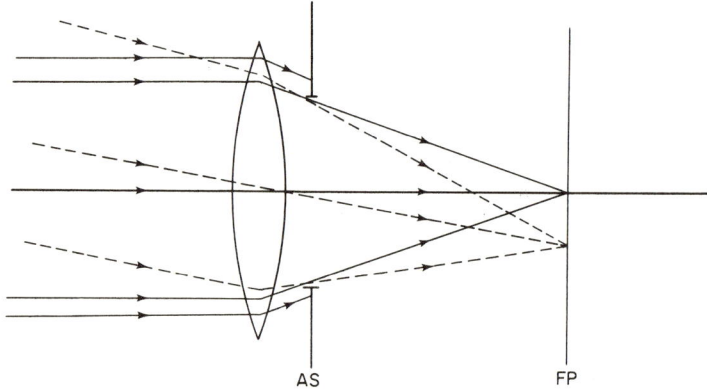

Fig. 6.1. An aperture stop AS for a single thin lens system.

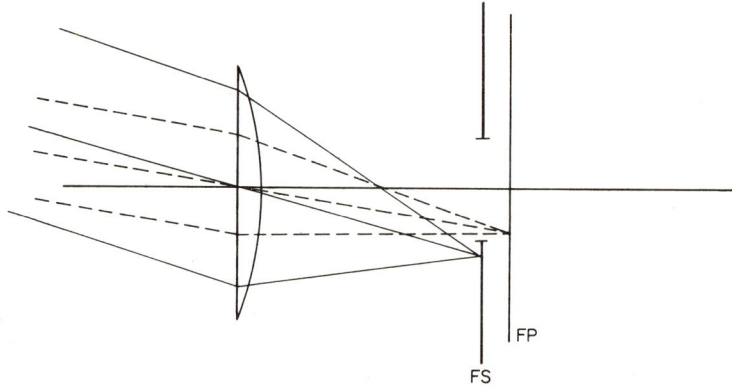

Fig. 6.2. A field stop FS for a single thin lens system.

finite size of the lenses which make up the system. Thus we see that in building an optical system we must take into account not only the focusing properties of the lenses but also the physical diameter of all the lenses in the system.

Let us consider an optical system made up of two subsystems A and B. We will assume that these systems are large enough so that the aperture stop AS determines which of the rays entering the system will contribute to image formation. Our system is illustrated in Fig. 6.3. AS' is the image of the stop AS formed by subsystem A. That is, AS' is the back projection of AS into object space. Since AS' and AS are conjugate, every ray that passes through AS' will also pass through AS. Likewise, AS'' is conjugate to AS in image space, and all rays passing through AS will also pass through AS''.

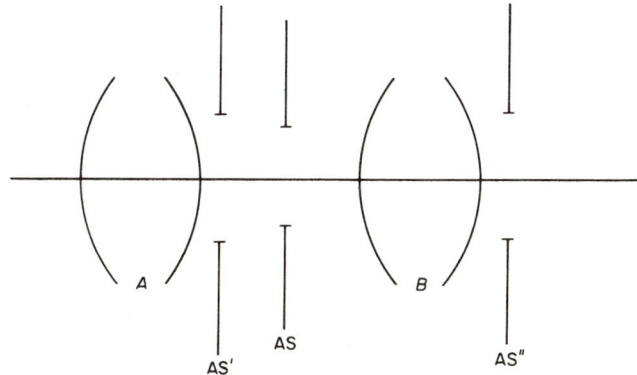

Fig. 6.3. An optical system consisting of two subsystems A and B. AS' is the image of AS in object space and AS'' is the image in image space.

Given a large number of elements for our optical system, each of which may function as an aperture, we project each in turn back into object space through the elements of the system that lie to its left. We then have a number of images of the apertures in object space (Fig. 6.4). For a given object point lying on the optic axis the limiting rays entering the system will be determined by the smallest aperture. This defines the *entrance pupil*. Clearly, the image AS_j' which functions as the entrance pupil will be determined by the position of the object being examined. In Fig. 6.4, AS_2' forms the entrance pupil for objects relatively close to the optical system while AS_3' forms the entrance pupil for those objects much farther removed from the system. If we now image this pupil through the entire system, we get the *exit pupil*. The exit pupil determines the limiting rays reaching the image in image space at the optic axis. The entrance and

66   CHAPTER 6   STOPS AND PUPILS; CHROMATIC ABERRATION

exit pupils determine the opening into the optical system when viewed from object space and image space, respectively.

Matrix formalism lends itself quite easily to the determination of the positions of the pupils of an optical system. We can see this by considering

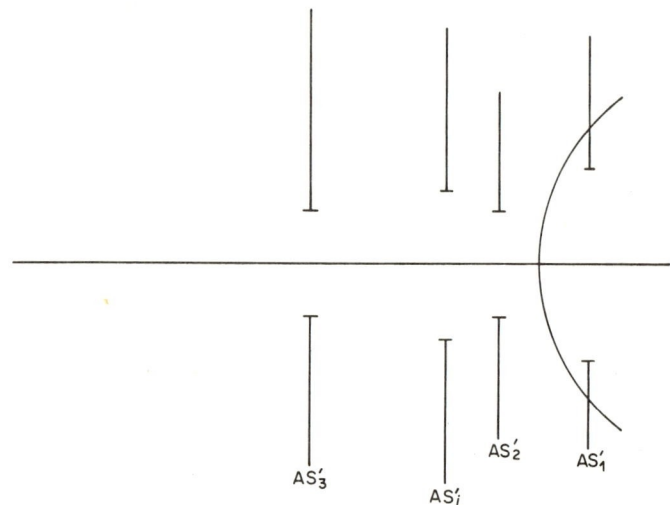

Fig. 6.4. The images of various stops in object space. AS_2' is the entrance pupil for distant objects.

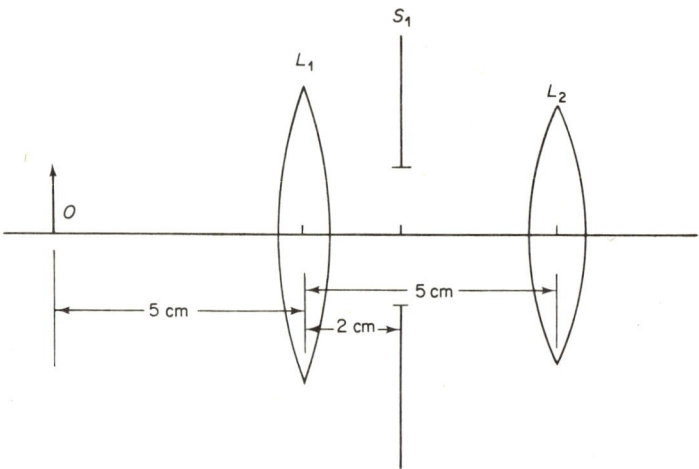

Fig. 6.5. An example of an optical system made up of a pair of thin lenses and a single stop. L_1: $f=4$ cm, aperture 6 cm. L_2: $f=8$ cm, aperture 5 cm. S_2 aperture 3 cm.

the example of an optical system made up of a pair of thin lenses and a single stop (Fig. 6.5). Since the focal lengths of the two thin lenses are given, we can write the system matrix as three matrices:

$$(S) = \begin{pmatrix} 1 & -\dfrac{1}{8} \\ 0 & 1 \end{pmatrix} \begin{pmatrix} 1 & 0 \\ 5 & 1 \end{pmatrix} \begin{pmatrix} 1 & -\dfrac{1}{4} \\ 0 & 1 \end{pmatrix}, \tag{6.1}$$

where we have assumed that the lenses are in air.

Since L_1 is the leading element of the optical system, it forms its own entrance pupil L_1'. The entrance pupil for S_1 is determined by the image of S_1 in L_1. For S_1 the object distance s_0 is $+2$ cm and the image position is found in the usual way by using Eq. (3.9) and the matrix for L_1:

$$s = \frac{0 + 1(+2)}{1 + \dfrac{1}{4}(+2)} = \frac{4}{3} \text{ cm}, \tag{6.2}$$

and the magnification using Eq. (3.10a) is given by

$$\mu = 1 - \frac{1}{4}\left(\frac{4}{3}\right) = \frac{2}{3},$$

so that the aperture S_1' due to S_1 is $\tfrac{2}{3} \times 3$ cm or 2 cm in diameter. Both L_1' and S_1' are entered in Fig. 6.6 where we show only the positions and sizes of the images of the apertures, and the position of the elements of the system are indicated as pips on the optic axis.

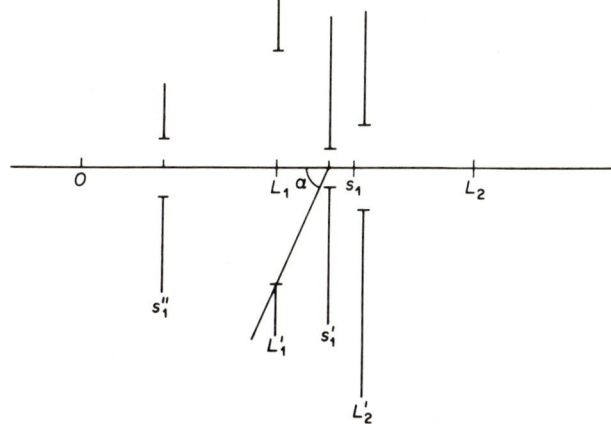

Fig. 6.6. The entrance apertures L_1' and L_2', the entrance pupil S_1' and the exit pupil S_1'' for the example in Fig. 6.5. The pips on the optic axis indicate the position of the lenses and the stop S_1.

CHAPTER 6 STOPS AND PUPILS; CHROMATIC ABERRATION

Finally we find the entrance pupil L_2' due to L_2 by imaging L_2 in L_1 and proceeding as with S_1. We get

$$s = \frac{0+1(5)}{1+\frac{1}{4}(5)} = \frac{20}{9} \text{ cm}$$

and

$$\mu = 1 - \frac{1}{4}\left(\frac{20}{9}\right) = \frac{4}{9},$$

so that the opening in the entrance aperture due to L_2 is $(4/9)5 = 20/9$ cm. Again, L_2' is shown in Fig. 6.6.

The entrance pupil in this case is determined by the stop S_1. In this case the result is obvious by inspection, but in cases where the result is not so clear-cut it is necessary to determine the angle subtended at O by each of the images, in which case the entrance pupil is that aperture which subtends the smallest angle at O.

In order to find the exit pupil S_1'', we now image S_1' through the entire optical system. In order to do this we first generate the system matrix (6.1):

$$(S) = \begin{pmatrix} \frac{3}{8} & -\frac{7}{32} \\ 5 & -\frac{1}{4} \end{pmatrix}, \tag{6.3}$$

and the Gaussian elements are given by

$$a = \frac{7}{32}$$

$$b = \frac{3}{8}$$

$$c = -\frac{1}{4}$$

$$d = -5.$$

Using these Gaussian elements to image $S_1'(s_0 = +4/3)$, we get

$$s = \frac{-5 + \left(-\frac{1}{4}\right)\left(\frac{4}{3}\right)}{\frac{3}{8} + \left(\frac{7}{32}\right)\left(\frac{4}{3}\right)} = -8 \text{ cm},$$

and the magnification is given by

$$\mu = -\frac{1}{4} - \left(\frac{7}{32}\right)(-8) = 1.5,$$

so that the physical size of the exit pupil is (1.5)2=3.0 cm. (Note that the diameter of the entrance pupil is 1 cm, not 3 cm!) S_1'' is shown in Fig. 6.6, 8 cm to the *left* of L_2.

Once the aperture stop has been determined for a given situation as in the example above, it may be possible to replace the lenses in the system with smaller diameter lenses. For example, in Fig. 6.6, L_1' can be 1 cm in diameter and still not form the entrance pupil. Use of a lens 6 cm in diameter is overdesign.

Vignetting

We now want to deal with off-axis points in the object plane. The ray from an off-axis point that passes through the axial point in the entrance pupil is called the *principal ray*. If we look at points farther and farther off the axis in the object plane, we will occasionaly find the situation arising in which the principal ray goes through the edge of another aperture. Consider Fig. 6.7 where EP is the entrance pupil and A is another aperture of the system. As we take points in the object plane O farther and farther away from the axis, we eventually reach point O' where the principal ray touches A. Only a small part of the light from O' will enter the optical system and only part of the entrance pupil EP will be illuminated. When viewed through the exit pupil, we will see an image of O' such as

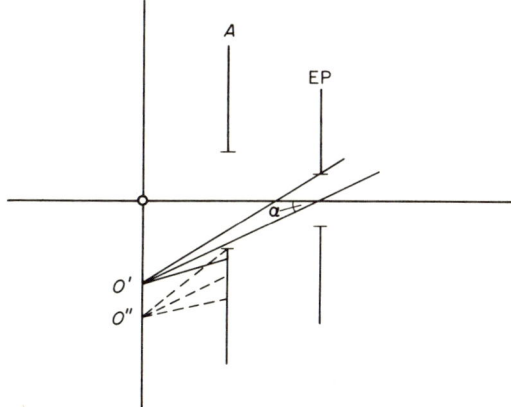

Fig. 6.7. Field of view and the vignetting of an object.

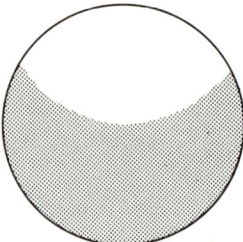

Fig. 6.8. Eyepiece view of a vignetted object.

is illustrated in Fig. 6.8. This partial illumination of the pupils is known as *vignetting*. Vignetting is most commonly encountered in optical systems with relatively large fields of view.

By extending the argument above, we can obviously reach a point O'' which is completely vignetted and no light from O'' will reach the exit pupil. We define the field of view of an optical system as that region in which the illumination at the edge in the image plane is one-half the central (axial) illumination. This can be approximated by taking the region in which A just passes the principal ray. A is, of course, the field stop. The field of view is generally specified by the angle α in Fig. 6.7.

If we go back to our earlier example and look at Fig. 6.6, S_1' defines the entrance pupil and L_1' is the field stop. We can find the field of view by finding α. Using our previous results we find

$$\tan \alpha = \frac{3}{\frac{4}{3}} = 2.25$$

and $\alpha = 66°$, a rather large field.

Chromatic Aberrations

Dispersion, the change in refractive index with changing wavelength, is a well-known phenomenon. Table 6.1 lists the various common Fraunhofer spectral wavelengths, and Table 6.2 gives the refractive indices for four of these lines. As can be seen from Table 6.2, the refractive index of glasses tends to increase with decreasing wavelength. As a result of this change in index with wavelength, both the refraction matrix and the translation matrix will vary with wavelength. We will find that red light and blue light will be focused at slightly different positions, thus affecting the quality of the image formed by the optical system (Fig. 6.9).

CHROMATIC ABERRATIONS

Table 6.1
FRAUNHOFER LINES

Wavelengths (Å)	Fraunhofer designation	Spectral source
4047	h	Hg
4341	G' (violet)	H
4359	g	Hg
4861	F (blue)	H
5461	e	Hg
5875	d	He
5893	D (yellow)	Na
6563	C (red)	H
7065	b	He

Table 6.2
REFRACTIVE INDICES OF FRAUNHOFER LINES

Medium	Refractive index			
	n_C	n_D	n_F	$n_{G'}$
Crown	1.49776	1.50000	1.50529	1.50937
Spectacle crown	1.52042	1.52300	1.52933	1.53435
Telescopic flint	1.52762	1.53050	1.53790	1.54379
Light flint	1.57208	1.57600	1.58606	1.59441
Dense flint	1.64357	1.64900	1.66270	1.67456

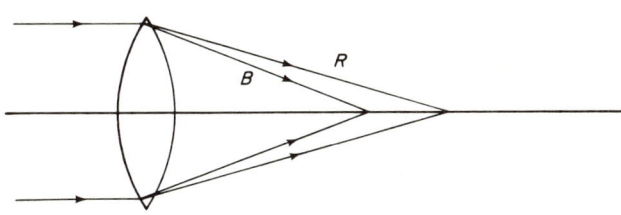

Fig. 6.9. Axial chromatic aberration showing the focal points for a blue (B) and red (R) ray.

Two effects are found. The longitudinal displacement of the image along the optic axis is called axial chromatic aberration, and the variation in image size with color is called lateral chromatic aberration (Fig. 6.10). It is possible within the framework of paraxial-ray theory to correct for these effects. The focal length may vary by about 2% with color over the spectrum for an uncorrected system, and with correction this may be lessened to a 0.2% variation in the same region.

We define the dispersive power of a medium as

$$\nu = \frac{n_F - n_C}{n_D - 1}, \tag{6.4}$$

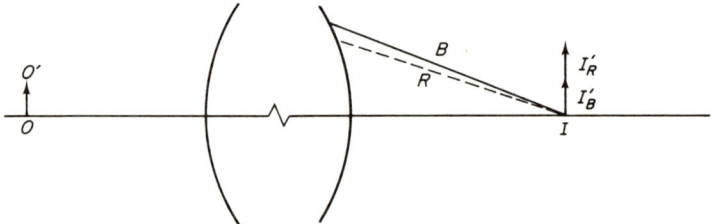

Fig. 6.10. Lateral chromatic aberration.

and we want to consider the effect of dispersion on an optical system. The Gaussian constant a in our general system matrix depends linearly on n, and for a thin lens system in air

$$a = (n-1)f(R_i),$$

where $f(R_i)$ is some function of the radii of curvature of the lens surfaces (Eq. 4.28).

The variation in power for a change in n is then given by

$$\delta a = \delta n f(R_i) \tag{6.5}$$

and

$$\frac{\delta a}{a} = \nu, \tag{6.6}$$

where we have chosen δn as $n_F - n_C$.

We now want to try to construct a lens for which δa is zero. Initially we will investigate the use of two thin lenses with power a_1 and a_2 where the lenses are in contact. Equation (4.25) shows that for this situation

$$a = a_1 + a_2, \tag{6.7}$$

and with Eq. (6.6) we see that

$$\delta a = a_1 \nu_1 + a_2 \nu_2. \tag{6.8}$$

If we set $\delta a = 0$ and a equal to the required power of the lens system, we

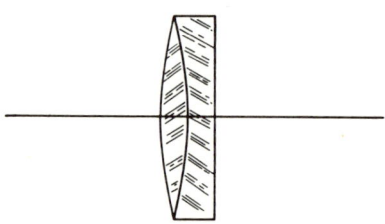

Fig. 6.11. An achromatic cemented doublet.

can satisfy the conditions (6.7) and (6.8) by using lens materials with different dispersive powers. For example, ν for dense flint is about $1/30$ and ν for crown glasses is about $1/60$. If we solve (6.7) and (6.8) simultaneously for the condition $\delta a = 0$, we get

$$a_2 = \frac{-a\left(\dfrac{\nu_1}{\nu_2}\right)}{1 - \dfrac{\nu_1}{\nu_2}}$$

$$a_1 = \frac{a}{1 - \dfrac{\nu_1}{\nu_2}},$$
(6.9)

and since $\nu_1 \neq \nu_2$, either a_1 or a_2 will be negative.

The typical achromatic lens is made by cementing a positive and a negative lens together as illustrated in Fig. 6.11. The focal length as a function of wavelength is shown in Fig. 6.12. Note that F and C light must have the same focal length as a result of the conditions we have fixed above.

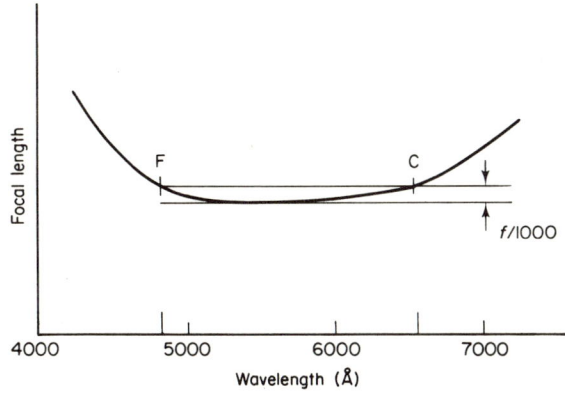

Fig. 6.12. The focal length as a function of wavelength for the achromatic doublet in Fig. 6.11.

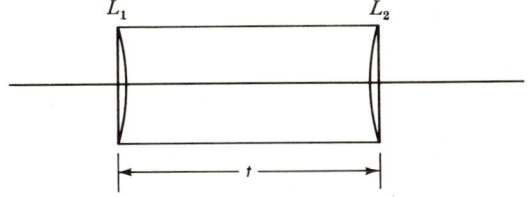

Fig. 6.13. The Ramsden eyepiece. $L_1: a = a_1$. $L_2: a = a_2$.

We also want to investigate the color correction of a typical eyepiece used in optical instruments. This eyepiece, known as the Ramsden eyepiece, is illustrated in Fig. 6.13. We use two thin lenses with powers a_1 and a_2 and specify that they are constructed of the same glass and separated in air by a distance t. The system matrix then becomes

$$(S) = \begin{pmatrix} 1 & -a_2 \\ 0 & 1 \end{pmatrix} \begin{pmatrix} 1 & 0 \\ t & 1 \end{pmatrix} \begin{pmatrix} 1 & -a_1 \\ 0 & 1 \end{pmatrix},$$

so that

$$(S) = \begin{pmatrix} 1 - a_2 t & -(a_1 + a_2 - a_1 a_2 t) \\ t & 1 - a_1 t \end{pmatrix}, \tag{6.10}$$

and the power of the system is given by

$$a = a_1 + a_2 - a_1 a_2 t. \tag{6.11}$$

Proceeding as before

$$\delta a = \delta a_1 + \delta a_2 - t \delta(a_1 a_2) = 0, \tag{6.12}$$

and using the rules of differentiation

$$\delta a_1 + \delta a_2 - t a_1 \delta a_2 - t a_2 \delta a_1 = 0. \tag{6.13}$$

If we now substitute for the a_i's using Eq. (6.6) where $\nu_1 = \nu_2 = \nu$ since we have assumed both lenses are constructed from the same material, we get

$$a_1 \nu + a_2 \nu = t(a_1 a_2 \nu + a_1 a_2 \nu),$$

which simplifies to

$$t = \frac{1}{2} \left(\frac{a_1 + a_2}{a_1 a_2} \right) = \frac{1}{2} \left(\frac{1}{a_1} + \frac{1}{a_2} \right), \tag{6.14}$$

and since $1/a_1 = f_1$ and $1/a_2 = f_2$ we have

$$t = \frac{1}{2}(f_1 + f_2), \tag{6.15}$$

and we find that the eyepiece (Fig. 6.13) has no longitudinal chromatic aberration when the lenses are separated by half the sum of their focal lengths. This allows us to use simple rather than doublet lenses in the construction of eyepieces with substantial savings in cost.

We have treated only the correction for longitudinal chromatic aberrations in this section. Lateral chromatic aberrations can also be corrected by using, for example, pairs of achromatic doublets and then adjusting their spacing for constant magnification. We will not investigate this problem in more detail here.

Problems

6.1 A thin lens with a focal length of -10 cm and an aperture of 4 cm has a 2-cm stop located 4 cm in front of it. An object 4 cm high is located 12 cm in front of the lens. Locate the exit pupil and the image and characterize the image.

6.2 A thin lens L_1 ($f=6$ cm, aperture 3 cm) is located 5 cm in front of L_2 ($f=-10$ cm, aperture 5 cm). An object 1 cm high is located 20 cm in front of L_1. A stop, 3 cm diameter, is located 5 cm in front of L_1. Find the pupils, the image, and characterize the image.

6.3 An exit pupil 4 cm in diameter is located 10 cm in front of a spherical mirror $r=-16$ cm. Find the entrance pupil.

6.4 A thin lens $f=12$ cm, 4 cm in diameter, is placed midway between the eye and an object 10 cm from the eye. What is the largest object that can be viewed?

6.5 A cemented achromatic doublet is to be made of light flint and crown glasses. The crown glass is to be equiconvex and the focal length of the combination is to be 20 cm. Give the lens parameters of the flint lens.

6.6 Repeat Problem 6.5 for a 12-cm focal length combination in which the flint lens is plano-concave.

Chapter 7

Geometric Aberrations

All the systems considered thus far have been treated in the limit of the paraxial-ray assumption. We will now discuss the deviations that occur when this assumption breaks down. In Chapter 2 we showed that the paraxial-ray assumption is equivalent to writing Snell's law in the form

$$n_1 \beta_1 = n_2 \beta_2, \qquad (7.1)$$

rather than as

$$n_1 \sin \beta_1 = n_2 \sin \beta_2. \qquad (7.2)$$

If we expand $\sin \beta$ in a Taylor expansion we get

$$\sin \beta = \beta - \frac{\beta^3}{3!} + \frac{\beta^5}{5!} - \frac{\beta^7}{7!} + \cdots, \qquad (7.3)$$

and we can see that Eq. (7.1) is obtained from Eq. (7.2) by taking only the leading term of the expansion (7.3).

If we develop our theory using the first two terms of (7.3), we find nonlinear terms of the third order in the various image-object expressions. Experimentally the images will no longer be sharp. This occurs, for example, when the angles involved in image formation are large, i.e., for objects far off the optic axis, or for points quite close to the vertices of our optical system. These effects are known as image aberrations or geometric aberrations and we limit ourselves here to third-order aberrations —those arising when the cubic term in the sine expansion becomes important.

The complete treatment of even third-order aberrations is beyond the scope of this book. We can, however, present certain physical arguments which will show the nature of the image defects and then describe the effect as seen in real systems.

We have found within the limits of the paraxial-ray assumption that for each point P in some plane in object space, the object plane, there is a corresponding point P' on a conjugate plane in image space, the image plane (Fig. 7.1). The implication is that any ray from the object passing through the system and reaching the image plane will be of constant optical length. In fact, if we look at a system where we take only P and P' and the entrance and exit pupils (Fig. 7.2), we see that the bundle of rays leaving P and accepted by the system will have spherical surfaces at a fixed distance from P as long as the medium is uniform (no surfaces intercepted). Likewise the exit bundle of rays leaving the exit pupil and moving to P' should have spherical surfaces about P' as a center at a constant length from P' once the rays are out of the inhomogeneous part of the system and into a medium of constant index of refraction.

We can interpret these surfaces in image space in terms of the time of transit. Since the index of refraction is related to the velocity of propagation of a light ray, surfaces such as DD', EE', and FF' should be surfaces separated from the center P' by equal times.

In real optical systems we find that the equal time surfaces centered about P' deviate from the ideal spherical surface we would find in the Gaussian approximation. Figure 7.3 gives an example of one situation which may arise. We find in Fig. 7.3, rather than the ideal spherical

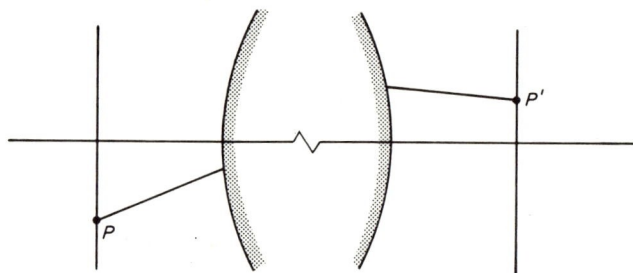

Fig. 7.1. Conjugate planes containing an object point P and an image point P'. P lies in the object plane and P' in the image plane.

GEOMETRIC ABERRATIONS

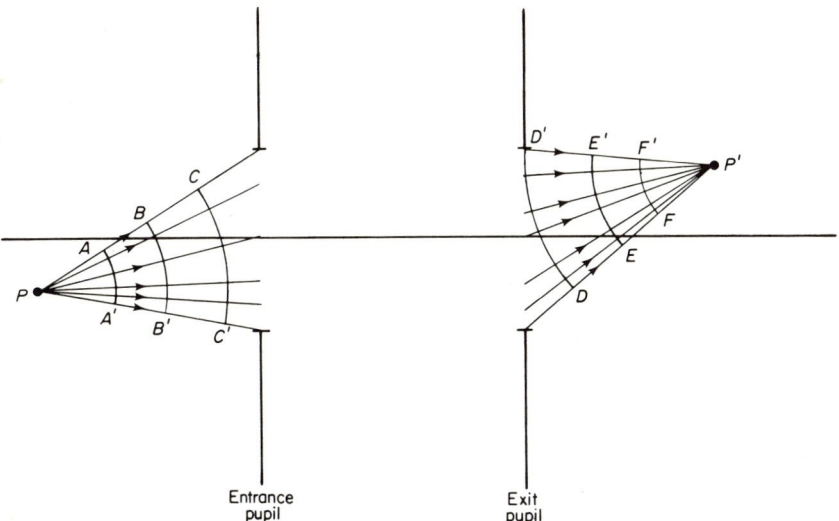

Fig. 7.2. The entrance and exit pupils of a generalized optical system. A number of ideal spherical surfaces at constant separation from P and P' are shown.

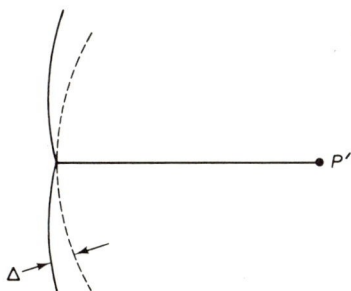

Fig. 7.3. The deviation from ideal in a system containing aberrations.

surface indicated by the dotted lines, the real surface deviating from this ideal by an amount Δ which is a function of the position on the surface. We will continue to assume a cylindrical symmetry about the optic axis of the system so that we can continue to use cross-sectional diagrams for the systems.

One way of treating aberration is to expand Δ in a power series. This generally leads to a very complicated algebraic expression. However, it is possible to express Δ in terms of five constants

$$\Delta = f(S_1, S_2, S_3, S_4, S_5), \tag{7.4}$$

known as the Seidel sums. If all five S_i were zero, then images would be free of all defects. No real optical system satisfies all these conditions at any one time. $\varDelta$ equal to zero is the ideal Gaussian system. The disappearance of various of the Seidel sums corresponds to the disappearance of certain aberrations, and we will discuss separately each of these five aberrations:

>Spherical aberration
>Coma
>Astigmatism
>Curvature of the field
>Distortion

Spherical Aberration

If $S_1=0$, then no spherical aberration occurs, i.e., the marginal rays (those passing near the edges of the optical system) focus at the same point as the central paraxial rays. When $S_1 \neq 0$ we have a situation similar to that illustrated in Fig. 7.4. Marginal rays tend to be more sharply focused than central rays, and the distance FM is a measure of the *longitudinal spherical aberration*. This is a function of h^2. We also have a lateral spherical aberration which is a measure of the deviation of the marginal rays from a point focus in the paraxial focal plane.

Figure 7.5 shows a graph of focal length versus h for a thin lens. The ideal curve would be a vertical line such as the dotted line in Fig. 7.5. We see that the curve approximates a parabola, which it must because

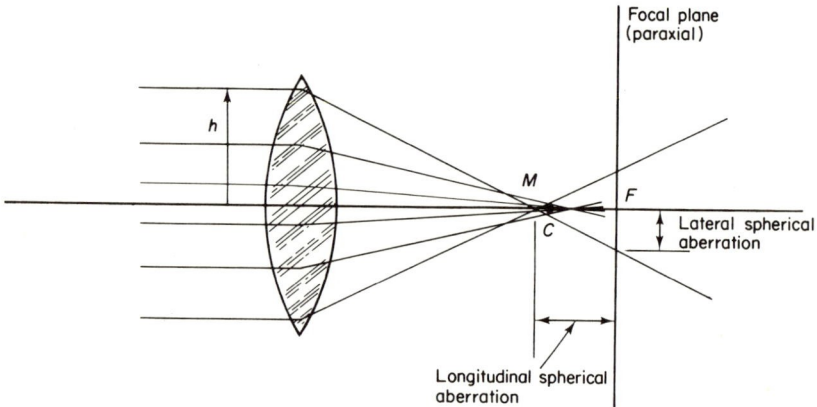

Fig. 7.4. Spherical aberration showing the difference in focusing between the central rays. C labels the circle of least confusion.

SPHERICAL ABERRATION

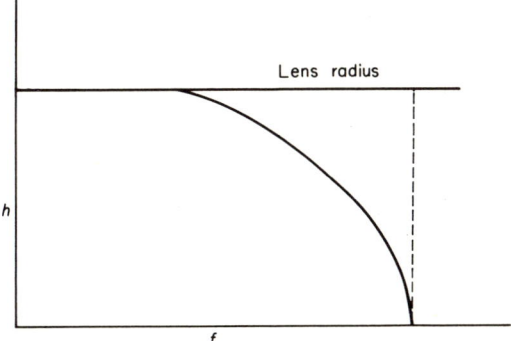

Fig. 7.5. Focal length f versus the displacement h of the object from the optic axis.

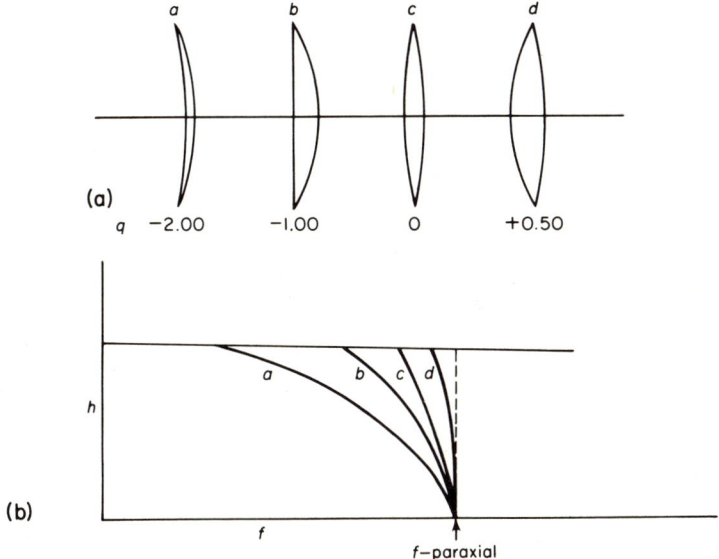

Fig. 7.6. (a) Four lenses with the same focal length but varying shape factor q. (b) h versus f curves for the lenses in (a).

of the h^2 dependence of the focusing of the marginal rays. If in Fig. 7.4 we move a screen from F to M, the image we observe on the screen will be a circle rather than a point, and at some position C the size of this circle will be minimal. This is known as the *circle of least confusion*.

Spherical aberration cannot be eliminated for a single spherical lens but it can be removed with two spherical lenses of opposite sign. It is possible, however, to minimize the spherical aberration with a single lens of fixed focal length by adjustment of the shape factor q:

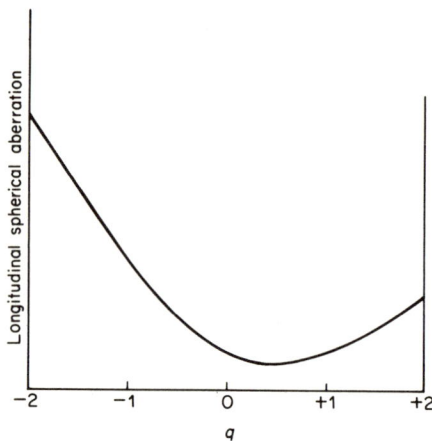

Fig. 7.7. Longitudinal spherical aberration versus q for a typical thin lens.

$$q = \frac{r_2 + r_1}{r_2 - r_1}. \tag{7.5}$$

Figure 7.6a illustrates four lenses, all with focal length f and diameter d, but with varying shape factors. The h versus f curves for these lenses are given in Fig. 7.6b. Here we see that the "bending" of the lens can be used to minimize spherical aberration. Figure 7.7 represents the curve of q versus longitudinal spherical aberration for a typical thin lens. We see that for some value of the shape factor q the aberration is minimized. It does not disappear, however, which is consistent with the statement that a single thin lens cannot have $S_1 = 0$ if the lens is spherical. Spherical aberration is eliminated for nonspherical lenses by a process known as aspherizing, in which the curvature of the lens is made to vary with the distance from the center h.

It should be clear that the system matrix as we have developed it in the earlier chapters is independent of q. For any one of the lenses illustrated in Fig. 7.6a, we can write the system matrix

$$(S) = \begin{pmatrix} 1 & \dfrac{1}{f} \\ 0 & 1 \end{pmatrix}$$

since they are all thin lenses. Thus we see that in order to be able to handle aberrations within the matrix theory we must either increase the dimensionality of the matrices or we must introduce new variables into the matrix theory. The general development of such a theory is sketched in Appendix 2.*

---

* The reader is also referred to the book by Brouwer for further discussion of this subject. See the Bibliography at the end of this book.

Coma

The next aberration which we will consider will be coma. In order for a system to be free of coma, both S_1 and S_2 must be zero. The name of the effect derives from the shape of the image of a point source, and we deal here with objects located near but not on the optic axis (see Fig. 7.8). Even though a system can bring all the axial points into good focus (no spherical aberration), off-axis points must be corrected for separately.

If we look at the plot of shape factor versus coma C_T for a thin lens and measured on some arbitrary scale (Fig. 7.9) we see that the zero of coma and the minimum of spherical aberration lie close to the same value of q. Unlike spherical aberration, coma can be completely elimi-

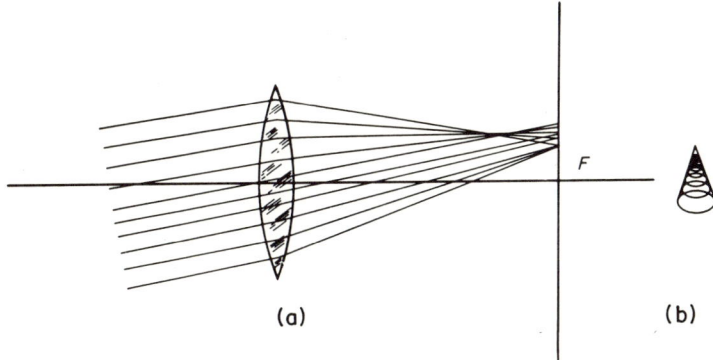

Fig. 7.8. (a) Formation of the comatic image. (b) The comatic image.

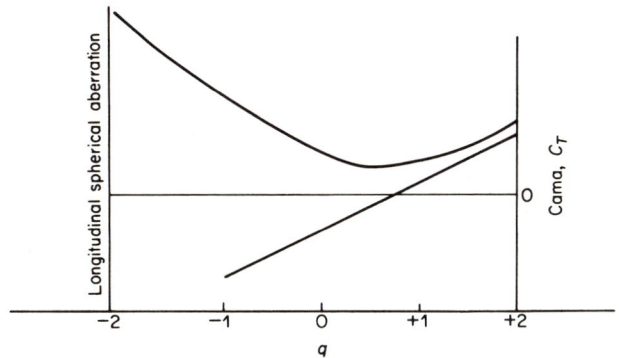

Fig. 7.9. Longitudinal spherical aberration and coma versus q for a thin lens showing the near coincidence of the minimum in the spherical aberration and the zero of coma.

nated in a thin lens by proper choice of the shape factor. This value of q will often be so close to the minimum value of the spherical aberration that the spherical aberration will be minimized also.

When the optical system is such that $S_1 = S_2 = 0$, the system is said to be aplanatic. Except for extremely rare cases, such systems are made up of aspherical surfaces and thus are generally difficult to construct.

Astigmatism

Astigmatism is the term applied to the image aberration arising as a result of object points lying some distance away from the optic axis. If $S_1 = S_2 = 0$, we find that the optical system generates point images from point objects on or near the optic axis; thus, there will be no coma or spherical aberration. When $S_3 \neq 0$, we have the situation illustrated in Fig. 7.10. The tangential rays, those rays lying in the plane of the object OO', and the optic axis are focused at TT', while the sagittal rays, those lying in the plane normal to the tangential plane, are focused at SS'.

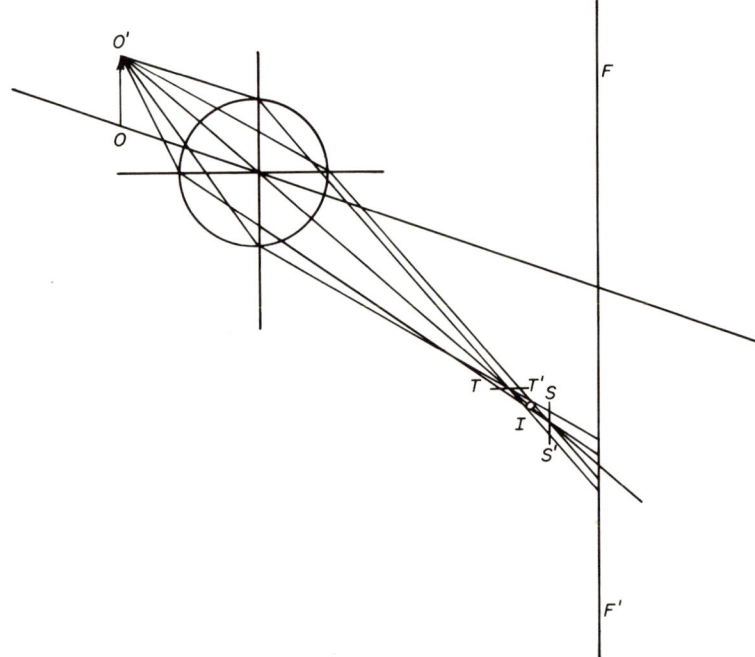

Fig. 7.10. Formation of the astigmatic image.

TT' and SS' are called focal lines, and the focal lines are normal to the tangential and sagittal planes.

At some point I intermediate between SS' and TT' the image of a point source at O' will be circular, and this intermediate point I constitutes the circle of least confusion for the case of an astigmatic system. Astigmatism is measured by the distance between SS' and TT' measured along the principal ray. Clearly the astigmatic difference depends on the angle the principal ray makes with the optic axis. Figure 7.11 represents one way of looking at this problem. For a given direction of the principal ray we find its intersection with a paraboloid of revolution representing the sagittal focal surface. The optical system is schematically represented here by LL', and when TT' lies between LL' and SS', as in this figure, the astigmatism is said to be positive.

Astigmatism is insensitive to changes in the shape factor q and is best reduced by the introduction of stops or in some cases by altering the

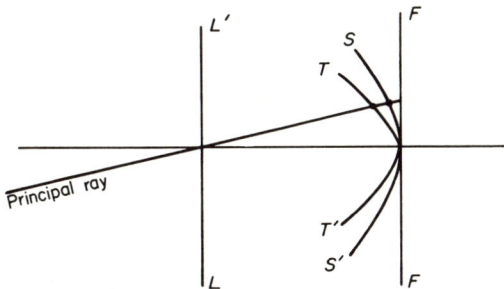

Fig. 7.11. The tangential and sagittal focal points for a principal ray moving in a given direction.

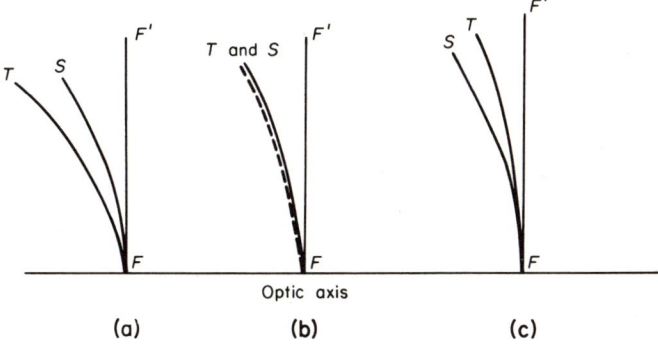

Fig. 7.12. (a) Astigmatism of a single thin lens. (b) Astigmatism corrected by the addition of a stop in the appropriate position. (c) Effect of incorrect placement of the stop, giving negative astigmatism.

spacing of the elements within the system. This reduces the curvature of the focal paraboloids. For example, Fig. 7.12a shows the situation for astigmatism in a single lens. With the introduction of a stop in the proper position, S and T can be made to coincide and therefore the system can be made free of astigmatism as in Fig. 7.12b. Further alteration of the spacing can lead to negative astigmatism. Point images of point objects far off the optic axis will only be formed for the situation shown in Fig. 7.12b.

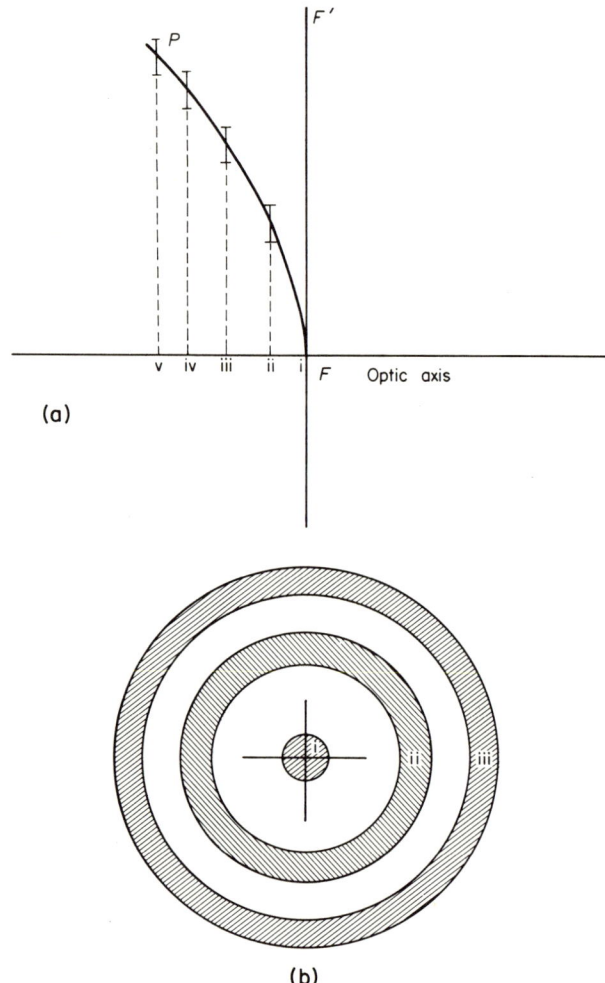

Fig. 7.13. (a) The Petzval surface. (b) The region of sharp image formation for screens placed as indicated in (a).

Curvature of the Field

If we now have corrected our optical system so that $S_1 = S_2 = S_3 = 0$, we find that the surface over which point images are formed is the surface of rotation generated by the hemiparabola in Fig. 7.12b. This surface on which point images are formed is known as the Petzval surface. If we now try to form an image on a flat plate such as a photographic plate, only some small circular region will be in sharp focus. In Fig. 7.13b we illustrate this for some of the screen placements indicated in Fig. 7.13a.

A Petzval surface exists in theory for every optical system. It can be shown that, for a given distance from the optic axis, the separation of the Petzval surface from the tangential focal surface is always three times the separation from the sagittal surface. By proper adjustment of the astigmatism, one can bring the Petzval surface into such a position that a reasonably acceptable image can be formed over some reasonably sized region. This kind of compromise is common in less expensive cameras. The addition of a stop in front of the lens is also effective in reducing curvature of the field.

Distortion

Distortion, the final third-order aberration we will consider, may exist even if the first four aberrations are corrected. Distortion arises when the lateral magnification over the field is nonuniform. A pinhole camera with no lens is completely free of distortion since straight lines join all object and image points. The introduction of an optical system can lead to distortions of the image as illustrated in Fig. 7.14. The square object (a) will have barrel distortion (b) if the magnification decreases radially

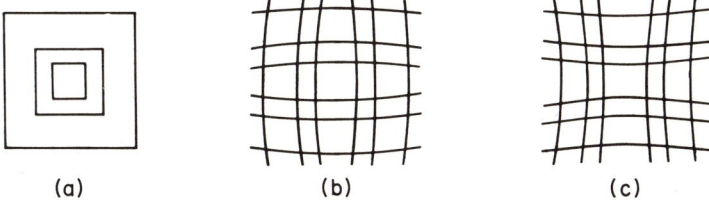

Fig. 7.14. (a) A square object. (b) The image formed of (a) when the system has barrel distortion. (c) The image formed of (a) when the system has pincushion distortion.

from the optic axis and pincushion distortion (c) if the magnification increases radially.

The single thin lens is generally free of distortion; however, the introduction of a stop at some point other than the lens introduces distortion into the system. Stops placed before the lens generally lead to barrel distortion, while stops placed behind the lens generally give pincushion distortion. To eliminate distortion in camera lenses, the lens combination is generally made up of a series of elements symmetric about a central stop.

Higher-Order Aberrations

Inclusion of the fifth-order term $\beta^5/5!$ in our theory would lead to still more problems. For example, the spherical aberration term would be of the form $ah^2 + bh^4$. For small values of h the fifth-order term is generally negligible, and where aberration corrections are made these are usually exclusively for the third-order terms. When fifth-order corrections are used, only those corresponding to fifth-order spherical aberration are treated in general. Fifth-order spherical aberration leads to images of point sources in the form or combination of the forms illustrated in Fig. 7.15.

Fig. 7.15. Images formed of point sources by a system with fifth-order spherical aberration.

Chapter 8

Optical Instruments

In this chapter we discuss a number of common optical instruments within the paraxial-ray theory as developed in earlier chapters. In effect, this chapter represents the application of the ideas presented earlier and should, in addition, serve to illustrate the usefulness of the theory.

The Simple Magnifier

The simple magnifier, sometimes called a simple microscope, is a single lens or a lens combination used to form an enlarged virtual image of a real object. The eye can focus on objects from infinity in to a nominal 250 mm known as the near point, the closest point at which an average eye can form an image. This region from infinity to 250 mm is known as the range of accomodation, and the image formed by the simple magnifier must lie within this range.

The visual magnification achieved by use of some instrument is fixed by first considering the object to be at the near point, in which case the angle subtended at the eye is

CHAPTER 8 OPTICAL INSTRUMENTS

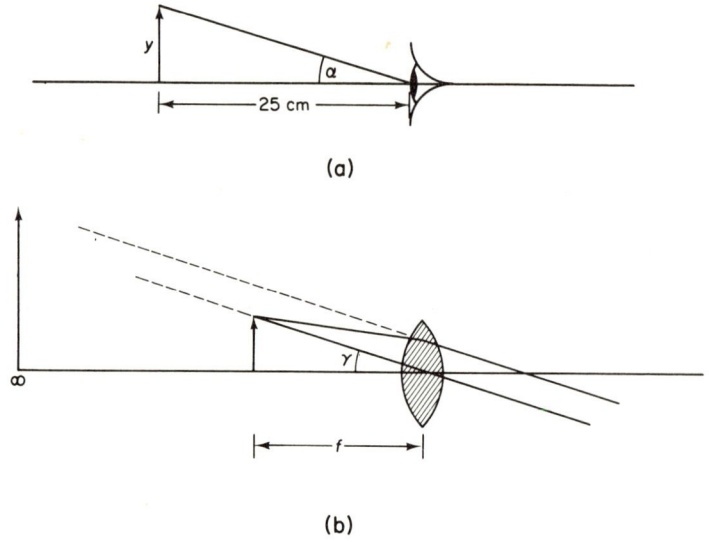

Fig. 8.1. (a) The angle α subtended at the eye by an object at the near point. (b) The effect of a simple magnifier.

$$\tan \alpha = \frac{y}{25}, \tag{8.1}$$

where y is the height of the object (Fig. 8.1a). With the aid of the magnifier this object can form a virtual image at infinity if the object is at the focal point of the magnifier, in which case the angle subtended is (Fig. 8.1b)

$$\tan \gamma = \frac{y}{f}. \tag{8.2}$$

The ratio of these two quantities is the visual magnification, usually called V:

$$V = \frac{\tan \gamma}{\tan \alpha} = \frac{25}{f}, \tag{8.3}$$

where f is the focal length of the magnifier in centimeters. The visual magnification (often called the power) is generally expressed as $5\times$, $7\times$, etc. (read 5 power, 7 power), and for a simple magnifier of focal length 10 cm the power is $2.5\times$.

We can verify in the following way the fact that for a simple magnifier (Fig. 8.1a) an object at the focus forms a virtual image at infinity. The system matrix for the simple magnifier is given by

THE SIMPLE MAGNIFIER

$$(S) = \begin{pmatrix} 1 & -a \\ 0 & 1 \end{pmatrix}, \tag{8.4}$$

and the image-object relationship (Eq. 3.9) is:

$$\mathbf{s} = \frac{d + c\mathbf{s}_0}{b + a\mathbf{s}_0} = \frac{0 + s_0}{1 + as_0}. \tag{8.5}$$

For a virtual object at infinity $s \to -\infty$, and this can only occur if the numerator is negative and the denominator goes to zero. The geometry of the problem makes the numerator negative (object to the left of the system) and the denominator goes to zero if $\mathbf{s}_0 = -1/a$. But $1/a = f$, so we have established the fact that an object at the focal point of a simple magnifier forms a virtual image at infinity.

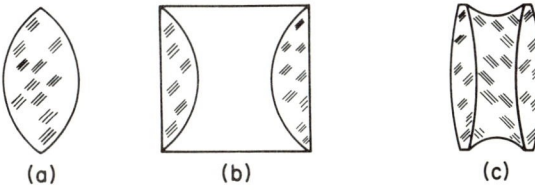

Fig. 8.2. Three simple magnifiers. (a) Simple. (b) Ramsden. (c) Cemented triplet.

Figure 8.2b illustrates a simple magnifier in the form of a Ramsden eyepiece. This has the advantage over the simple single lens by being corrected (Chapter 6). The system matrix for this combination is given in (6.10):

$$\begin{pmatrix} 1 - a_2 t & -(a_1 + a_2 - a_1 a_2 t) \\ t & 1 - a_1 t \end{pmatrix} = \begin{pmatrix} 0 & -a \\ \dfrac{1}{a} & 0 \end{pmatrix}, \tag{8.6}$$

where

$$t = \frac{1}{2}\left(\frac{1}{a_1} + \frac{1}{a_2}\right)$$

and we have assumed $a_1 = a_2$. This latter assumption is usually valid with simple magnifiers since they are constructed so that they are symmetric and thus can be used from either side. Figure 8.2 illustrates three basic forms. Note that in Fig. 8.2c the central lens has a decreased diameter, and this acts as a central stop and serves to minimize distortion over the field. The triplet combination of Fig. 8.2c is color corrected and has a relatively long working distance. The powers range up to 20×.

If we write the image-object relationship for matrix (8.6) we find that

$$s = \frac{-\frac{1}{a}}{as_0} = -\frac{1}{a^2 s_0}, \qquad (8.7)$$

and $s \to \infty$ only if $s_0 \to 0$, which means that the object would have to lie right at the vertex of the system to form an image at infinity. As a result, in the case of the simple magnifier, one usually sacrifices some of the color correction and makes

$$t < \frac{1}{2}\left(\frac{1}{a_1} + \frac{1}{a_2}\right)$$

so as to move the object position slightly away from the vertex to allow illumination of the object. The working distance is still very short.

Eyepieces

As we have previously pointed out, instruments for direct viewing of objects by the eye require us to form images within the range of accomodation. Thus a common feature in instruments such as telescopes and microscopes is the presence of an eyepiece designed to give a magnified virtual image within the range of accomodation. Figure 8.3 illustrates three of the most common eyepieces. The lens separation is such as to correct for chromatic aberration, and the curvatures and refractive indices of the glasses are chosen so as to minimize geometric aberration.

The Ramsden eyepiece, which we have already discussed, is generally made up of two lenses of equal focal length. They are usually separated by three-fourths the focal length of either lens slightly displaced from the optimum for correction of chromatic aberration, otherwise the images of the surfaces of the first lens (assumed thin) would coincide with the image of a distant object and reduce the image quality. We can find the principal focus of this combination by writing the system matrix for the eyepiece (in air) as

$$(S) = \begin{pmatrix} 1 & -\frac{1}{f_1} \\ 0 & 1 \end{pmatrix} \begin{pmatrix} 1 & 0 \\ \frac{3}{4}f_1 & 1 \end{pmatrix} \begin{pmatrix} 1 & -\frac{1}{f_1} \\ 0 & 1 \end{pmatrix}$$

and

$$(S) = \begin{pmatrix} \frac{1}{4} & -\frac{5}{4} \cdot \frac{1}{f_1} \\ \frac{3}{4} f_1 & \frac{1}{4} \end{pmatrix}. \qquad (8.8)$$

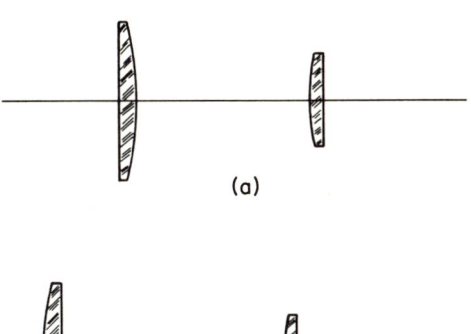

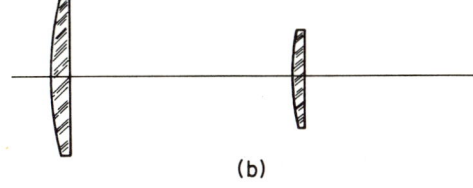

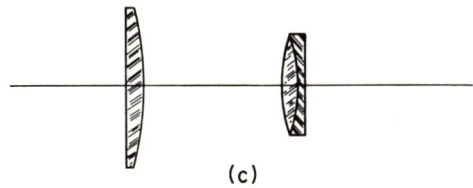

Fig. 8.3. Eyepieces. (a) Ramsden. (b) Huygens. (c) Kellner.

The principal focus can be gotten from the Gaussian elements of (8.8) by using Eq. (4.2) to fix the unit planes:

$$\mathbf{s}_u = s_u = \frac{c-1}{a} = \frac{-\frac{3}{4}}{\frac{5}{4}f_1} = -0.6f_1, \quad (8.9)$$

$$\mathbf{s}_{u0} = s_{u0} = \frac{1-b}{a} = 0.6f_1,$$

and $f_i = 1/a$ so that

$$f_i = 0.8f_1. \quad (8.10)$$

Figure 8.4a shows the positions of these cardinal points. A second advantage of having the focus before the lens system is that a stop or crosshairs or a recticle for scaling can be placed here, and both the scale and the object being viewed will be magnified by the entire eyepiece.

The Huygen's eyepiece (Fig. 8.3b) is generally constructed with the

94 CHAPTER 8 OPTICAL INSTRUMENTS

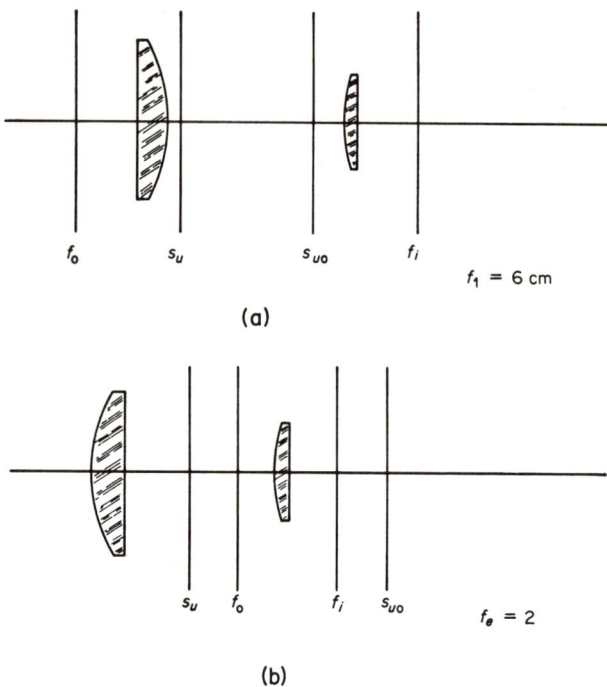

Fig. 8.4. The cardinal points of (a) a Ramsden and (b) a Huygens eyepiece.

first (field) lens having a focal length three times that of the second (eye) lens and with a separation between the lenses of half the sum of the focal lengths to achromatize the eyepiece. The system matrix is given by

$$(S) = \begin{pmatrix} 1 & -\dfrac{1}{f_e} \\ 0 & 1 \end{pmatrix} \begin{pmatrix} 1 & 0 \\ 2f_e & 1 \end{pmatrix} \begin{pmatrix} 1 & -\dfrac{1}{3f_e} \\ 0 & 1 \end{pmatrix}$$

so that

$$(S) = \begin{pmatrix} -1 & -\dfrac{2}{3f_e} \\ 2f_e & \dfrac{1}{3} \end{pmatrix}. \tag{8.11}$$

The unit planes are given in terms of the Gaussian elements of (8.11), and we find that

$$\mathbf{s}_u = s_u = \dfrac{c-1}{a} = -f_e,$$

$$\mathbf{s}_{u0} = s_{u0} = \dfrac{1-b}{a} = 3f_e, \tag{8.12}$$

and
$$f_i = \frac{1}{a} = \frac{3}{2}f_e. \tag{8.13}$$

The positions of these cardinal points are shown in Fig. 8.4b. With this eyepiece the incoming rays must converge at f_0 and form a virtual image for the eyepiece. The eyepiece therefore cannot be used as a simple magnifier as can the Ramsden eyepiece. In addition, cross-hairs or a reticle scale must be placed between the two lenses rather than in an external position, and therefore only the eye lens acts on the reticle, giving an unequal magnification relative to the object under study and chromatic aberration of the cross-hairs or reticle.

The Kellner eyepiece in Fig. 8.3c is designed to take advantage of the good properties of the Ramsden eyepiece and to eliminate the residual chromatic aberration caused by altering the lens spacing away from the the optimal. Use of a cemented doublet in the eye lens removes the remaining chromatic aberration as required.

Telescopes

The telescope is a device used to give enlarged images of distant objects. As we have seen in our discussion of simple magnifiers and eyepieces, the image presented to the eye is at infinity and thus we take a distant object which can be thought of as being at an infinite distance from the optical system and form an image again at infinity. This is a somewhat unusual situation and special conditions on the system matrix must be met. By using Eq. (4.8) with $p=q=\infty$, we find

$$\frac{1}{\infty} - \frac{1}{\infty} = a = 0 \tag{8.14}$$

and thus in a telescopic system we must have a system matrix

$$\begin{pmatrix} b & 0 \\ -d & c \end{pmatrix}. \tag{8.15}$$

In terms of the matrix equation for the telescopic system:

$$\begin{pmatrix} \lambda' \\ h' \end{pmatrix} = \begin{pmatrix} b & 0 \\ -d & c \end{pmatrix} \begin{pmatrix} \lambda \\ h \end{pmatrix},$$

so that
$$\lambda' = b\lambda$$

96 CHAPTER 8 OPTICAL INSTRUMENTS

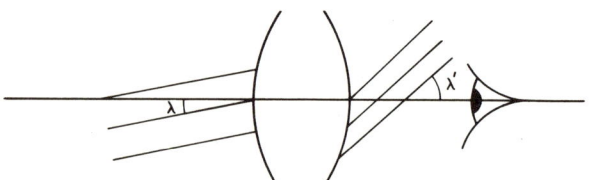

Fig. 8.5. Telescopic angular magnification.

and the Gaussian constant b is the *angular magnification* M of the system. The angular magnification is clearly the important quantity, as can be seen in Fig. 8.5 where the effect of the optical system is to change the angle subtended at the eye relative to the angle which would be subtended in the absence of the optical system.

The fact that $M=b$ immediately shows that a simple thin lens cannot act as a telescope since $b=1$ for all thin lenses. We can, however, produce a telescopic system from two thin lenses. This telescope, known as the Gallilean telescope, is the simplest telescope possible. The system matrix is given by

$$(S) = \begin{pmatrix} 1 & -a_2 \\ 0 & 1 \end{pmatrix}\begin{pmatrix} 1 & 0 \\ t & 1 \end{pmatrix}\begin{pmatrix} 1 & -a_1 \\ 0 & 1 \end{pmatrix}$$

so that

$$(S) = \begin{pmatrix} 1-a_2 t & -a_1-a_2+a_1 a_2 t \\ t & -a_1 t+1 \end{pmatrix}, \tag{8.16}$$

and setting $a=0$ we get

$$a_1+a_2-a_1 a_2 t = 0$$

$$t = \frac{a_1+a_2}{a_1 a_2} = \frac{1}{a_1}+\frac{1}{a_2} = f_1+f_2, \tag{8.17}$$

so that the separation of the lenses must be equal to the sum of their focal lengths. The angular magnification M is equal to b so that

$$M = 1-a_2 t = -\frac{a_2}{a_1}. \tag{8.18}$$

In order to make an enlarged image, $|a_2|>|a_1|$ and either a_1 or a_2 must be less than zero to obtain an erect image. The Gallilean telescope is shown in Fig. 8.6.

For many applications such as astronomical observation it is not necessary to have an erect image, and thus two positive lenses can be used in the construction of the system.

The pupils of a telescopic system are of primary importance. The edge of either lens in our system can serve as the entrance pupil. The first

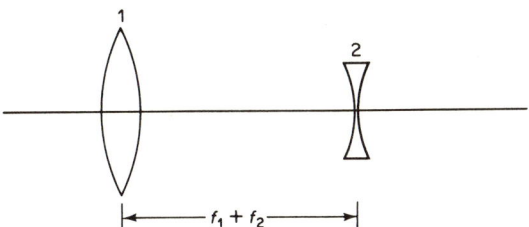

Fig. 8.6. The Gallilean telescope.

(objective) lens would act as the entrance pupil in its fixed position. The second (eyepiece) lens must be back-imaged through the objective into object space if we are to see its properties as a pupil. This is imaged through lens 1 which has a simple matrix given by

$$\begin{pmatrix} 1 & -a_1 \\ 0 & 1 \end{pmatrix} \tag{8.19}$$

and we find the image at

$$\mathbf{s}_2 = s_2 = \frac{0 + 1 \cdot t}{1 + a_1 t}, \tag{8.20}$$

where the subscript is written to indicate that this is the image of the second lens. This image has its diameter magnified μ times, where μ is given by

$$\mu = \frac{1}{1 + a_1 t}. \tag{8.21}$$

The general practice is to make the eyepiece lens large enough so that the objective lens acts as the entrance pupil. To find the diameter of the exit pupil we need to find the magnification of the telescope system which has the system matrix given by

$$(S) = \begin{pmatrix} 1 - a_2 t & 0 \\ t & 1 - a_1 t \end{pmatrix} \tag{8.22}$$

and

$$\mu = \frac{1}{(1 - a_2 t) + 0} = \frac{1}{M} \tag{8.23}$$

regardless of the position of the entrance pupil. The magnification is simply the ratio of the diameter of the objective lens to the exit pupil. Most monoculars and binoculars have stamped on their case the angular magnification and the diameter of the objective lens. For example, 7×35 means an angular magnification of $7\times$ and an objective lens 35 mm in diameter. By Eq. (8.23), this means that the exit pupil has a diameter of

5 mm, which is slightly larger that the diameter of the pupil of the eye in daylight. A larger exit pupil would not increase the light recieved by the eye. For night viewing an exit pupil up to 8 mm is useful since the diameter of the pupil of the eye increases in low light.

We have seen (Eq. 8.18) that the Gallilean telescope with a negative eyepiece gives an erect image while two positive lenses can give only an inverted image. If one wishes to construct an erect image using only posi-

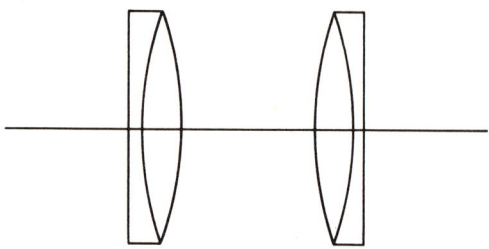

Fig. 8.7. Erecting lens for a terrestial telescope.

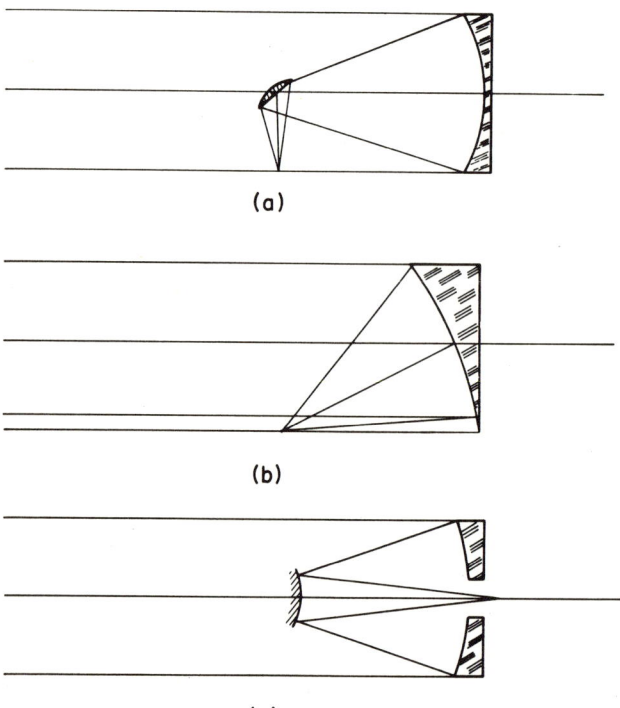

Fig. 8.8. Reflecting telescopes.

tive lenses, an erecting lens system can be inserted. This lens (Fig. 8.7) is inserted in such a way that it acts on the real image produced by the objective lens with a magnification of -1. This does, however, increase the length of the telescope by four times the focal length of the erecting lens.

In all but the simplest models, the objective lens would be an achromatic doublet, and one of the color corrected eyepieces illustrated in the previous section would be used as the eyepiece.

Particularly in astronomical applications where large light gathering is a necessary property of the system, mirrors are used as the objectives. These range up to 200 inches in diameter. Three common forms of reflecting telescope are shown in Fig. 8.8. The real image formed may be used directly for photography, or an eyepiece may be added for direct viewing. The theory developed in Chapter 5 is ideally suited for treating such cases.

The Microscope

The microscope, like the telescope, forms a virtual image of some object for direct viewing by the eye. The purpose is to achieve large magnification with minimal aberrations in the image. The microscope consists of two lens systems, an objective system which produces a real image of the object, and an eyepiece which images the real image formed by the first lens somewhere within the range of accommodation. Figure 8.9 is a diagram of the imaging process; we have not shown the actual lens system but simply the unit planes associated with the objective and with the eyepiece.

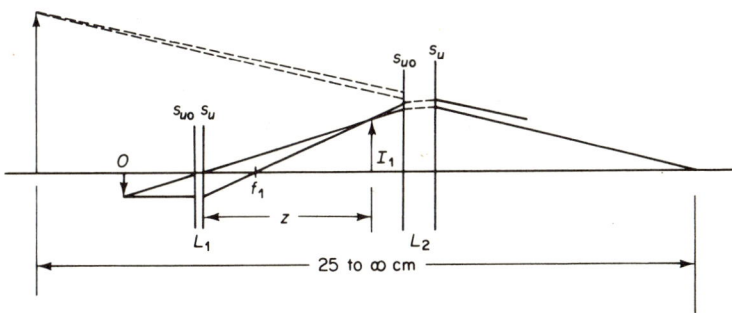

Fig. 8.9. The simple microscopic system showing only the unit planes of the lenses involved.

The objective forms a real image as shown in the figure, and the magnification of the object is just the linear magnification of the objective lens. The objective generally is a short-focal-length lens combination, and the object is placed near the focal point so that we can take the magnification to be $m_0 = -z/f_0$, where f_0 is the focal length of the positive objective. The quantity z is standardized by microscope manufacturers at 16 cm. Objective lenses, several of which are often placed on a turret on the microscope, are stamped with a magnification such as $50\times$ or $100\times$. The eyepiece lens converts the real image at the near point (250 mm) into a virtual image. The magnification here is just the visual magnification (Eq. 8.3), $M = 25/f_e$. This is usually stamped on the eyepiece, and many microscopes include interchangeable eyepieces. The over-all magnification is just $m_0 M$, and the user selects a suitable combination of lenses and eyepieces for the particular job at hand. Figure 8.10 shows some typical microscope objective lens combinations.

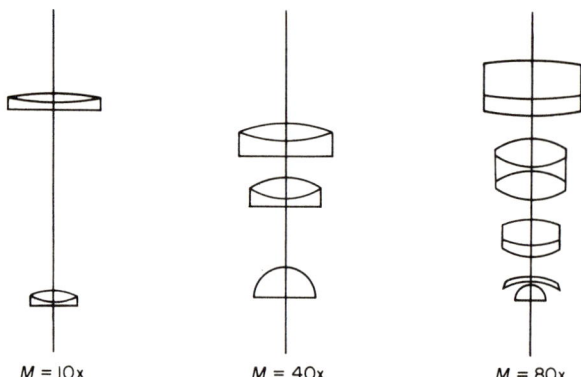

Fig. 8.10. Three examples of microscope objective.

The ability of the microscope to resolve small objects and also the illumination, i.e., the brightness of the image, depend on the numerical aperture of the system. The numerical aperture is the half-angle subtended at the object by the entrance pupil of the system (usually the first lens) and the index of the viewing medium:

$$\text{NA} = n \sin a, \qquad (8.24)$$

and this is as large as about 1.5. In order to increase the NA, the object may be immersed in oil. Certain objective lenses are designated as "oil immersion objectives."

The object is usually illuminated from below. In order to increase the

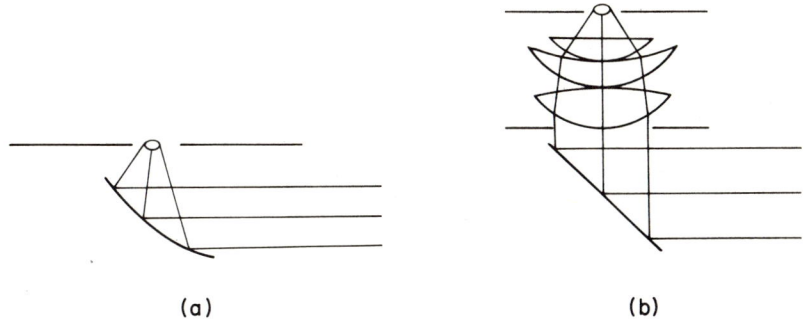

Fig. 8.11. Microscope condensers.

illumination the light source is condensed by a mirror or special condensing lens system as illustrated in Fig. 8.11. The light is *not* imaged at the object since that would lead to the superposition of an image of the sight source on the object when viewed through the microscope.

Cameras and Camera Lenses

The camera differs from many other optical instruments in that the image is formed on a recording film and therefore must be real rather than being formed as a virtual image for direct viewing with the eye. The camera body which holds the lens and film is a rigid structure constructed so as to allow accurate positioning of the lens relative to the film so that the film lies in the image plane. A direct viewing system is provided in conjunction with the recording optical system to help point and focus the system.

We limit our discussion of cameras to their most critical element, the lens. Figure 8.12 illustrates some basic lens types. Inexpensive cameras often use the single simple meniscus lens (Fig. 8.12a) which gives satisfactory results with small aperture and black and white film. Only astigmatism and coma are corrected, while spherical and chromatic aberration, distortion, and field curvature remain uncorrected. Figure 8.12b shows the basic form of most of todays better photographic lenses. It consists of a series of positive and negative lenses and is well corrected over the field for both geometric and chromatic aberrations. Figure 8.12c shows the basic Cooke triplet first developed for Cooke and Sons in England by H. D. Taylor. It is a well-corrected, relatively inexpensive lens usually used in some modification for less expensive cine cameras. Modifi-

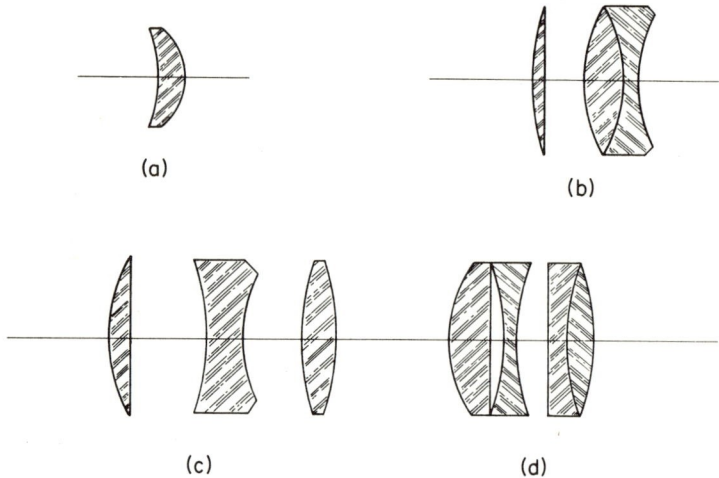

Fig. 8.12. Camera lenses. (a) Meniscus. (b) Basic compound. (c) Cooke triplet. (d) Tessar.

cations are often made by the introduction of an additional lens to correct spherical aberration which is relatively poorly corrected in the basic form. Figure 8.12d is the Tessar lens, a modification of the Cooke triplet, characterized by replacing one of the outer elements by a cemented doublet. This flattens the field and reduces the spherical aberration and in addition gives a relatively wide field of view (52°) at large aperture.

Camera lenses are characterized by their focal length and f-number. Since camera lenses are required to form real images of relatively distant objects, their focal lengths are positive. These are generally given by the lens manufacturer and stamped on the lens mounting. The f-number is the focal length divided by the diameter of the aperture stop and is, of course, a measure of the amount of light which passes through the system. Small f-numbers imply large openings, and for a given length of exposure on a standard film a small f-number implies that the lens will form a suitable image more quickly than for a large f-number, thus the f-number is sometimes referred to as the speed of the lens.

Table 8.1 gives the parameters for the Tessar lens. The system matrix can easily be found:

$$(S) = \begin{pmatrix} 0.8489 & -0.1968 \\ +1.3387 & 0.8675 \end{pmatrix}$$

$$|S| = 0.99988.$$

The focal length measured from s_u as given in Table 8.1 is 5.082 cm,

Table 8.1
Parameters for the Tessar Lens[a]

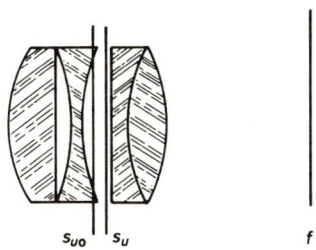

	n	t
—	1.000	—
$r_1 = 1.628$	1.6116	0.357
$r_2 = -27.57$	1.000	0.189
$r_3 = -3.457$	1.6053	0.081
$r_4 = 1.582$	1.000	0.325
$r_5 = \infty$	1.5123	0.217
$r_6 = 1.920$	1.616	0.396
$r_7 = -2.400$	1.000	—

[a] $S_{u0} = 0.7678$. $S_u = -0.6733$. $f = 1/a = 5.082$.

a rather typical value for a 35-mm still camera. The unit planes and the focal plane are shown in the diagram associated with Table 8.1.

We next consider a special lens system, often used with cameras, where the camera body allows for the interchange of lens systems. The telephoto system is used with distant objects to form images of sufficient size for recording on films. The size of the image depends on the magnification of the optical system, and for distant objects

$$\mu = \frac{1}{b + a\mathbf{s}_0} \approx \frac{1}{a\mathbf{s}_0}, \tag{8.25}$$

where we have taken $a\mathbf{s}_0 \gg b$. If we now express Eq. (8.25) in terms of the focal length (Eq. 4.11),

$$\mu \approx \frac{f}{\mathbf{s}_0}, \tag{8.26}$$

so we see that the magnification of distant objects is directly proportional to the focal length of the lens. An extremely long focal length lens would make the camera difficult to handle so a lens combination known as a telephoto lens has been developed to give long effective focal length without making the size of the lens extreme.

The simplest telephoto lens combination is that which combines a posi-

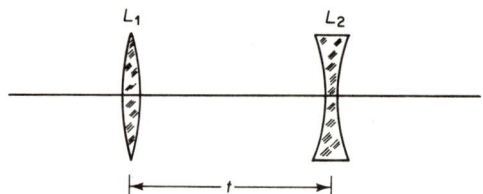

Fig. 8.13. A simple telephoto lens.

tive and a negative lens as illustrated in Fig. 8.13. We will assume the lenses to be thin and to have powers a_1 and $-a_2$, respectively, and to be separated by a distance t. The system matrix is given by

$$(S) = \begin{pmatrix} 1 & a_2 \\ 0 & 1 \end{pmatrix}\begin{pmatrix} 1 & 0 \\ t & 1 \end{pmatrix}\begin{pmatrix} 1 & -a_1 \\ 0 & 1 \end{pmatrix}$$

$$= \begin{pmatrix} 1+a_2t & -(a_1-a_2+a_1a_2t) \\ t & -a_1t+1 \end{pmatrix}. \quad (8.27)$$

The effect of combining these lenses has been to move s_u to a position well in front of L_1. We find that

$$s_u = \frac{c-1}{a} = \frac{-a_1t+1-1}{a_1-a_2+a_1a_2t} = -\frac{a_1t}{a_1-a_2+a_1a_2t}, \quad (8.28)$$

and the focal length measured from the rear vertex of the lens combination (L_2) is given by

$$s_u + \frac{1}{a} = \frac{c}{a} = \frac{1-a_1t}{a_1-a_2+a_1a_2t}. \quad (8.29)$$

Thus, for a fixed focal length, this combination has the effect of bringing the focal plane much closer to the back vertex of the system (L_2). A typical value would be about 5 cm.

This is, of course, the simplest telephoto combination but it demonstrates the principle involved in their construction. One would want the system to be color corrected and the geometric aberrations to be minimized in some way depending on the application. Pincushion distortion

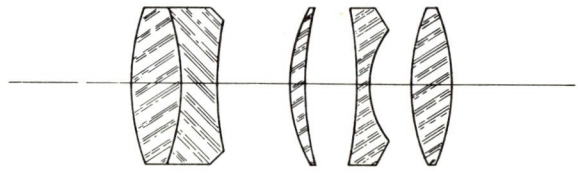

Fig. 8.14. Compound telephoto lens. $f/3.5$, 100 mm.

Projection Systems

Many cameras produce photographs in the form of transparent slides which are viewed by projecting them onto a screen. Figure 8.15 illustrates the construction of a typical slide projector. The source illuminates the condenser lens which forms an image of the source at the projection lens, completely filling the projection lens. This makes maximum use of the source since none of the light from it is wasted and maximum illumination of the slide is achieved. In addition, with the image of the source at the projection lens, it is impossible to image the source on the screen.

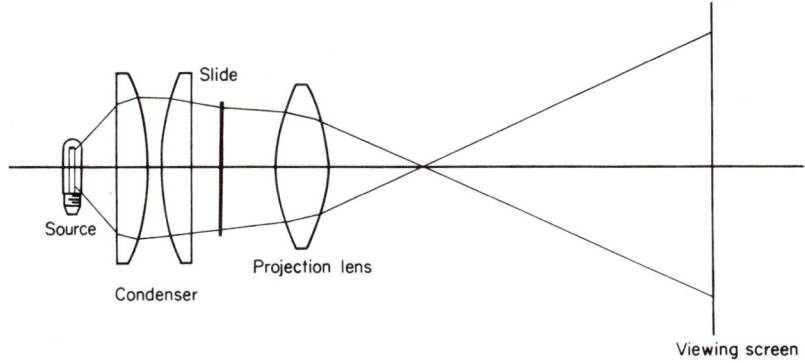

Fig. 8.15. A projection system.

The illuminated slide acts as the real objects for the projection lens which forms a real image of the slide on a viewing screen. Camera lenses in some modification are generally used with projectors and generally a small f-stop is sufficient for good projection.

Problems

8.1 Design and characterize the simple magnifier made of two identical plano-convex lenses with $f = 2$ cm.

8.2 A Ramsden eyepiece is made of two identical thin lenses $f = 2.0$ cm placed 1.5 cm apart. Characterize this eyepiece including the power.

8.3 Design a color-corrected Huygen's eyepiece with two lenses $f_1 = 2$ cm and $f_2 = 1$ cm. If L_1 has a 2-cm diameter, what should be the diameter of L_2 if L_1 is to serve as the entrance pupil?

8.4 How large is the exit pupil of an $8\times$ telescope having a 5-cm diameter objective? If the focal length of the objective is 15 cm, what is the eyepiece focal length?

8.5 Design a telescope around an achromatic objective $F = 30$ cm and diameter 5 cm. The exit pupil is to be 2 cm behind the eyepiece. Discuss both terrestial and astronomical types.

8.6 Find the system matrix and discuss the Cooke triplet in air characterized by:

Surface	r (cm)	t (cm)	n
1	4.0	6.0	1.6
2	−50	10	1.0
3	−5	1	1.6
4	40	11	1.0
5	24	6	1.6
6	−4	—	1.0

Appendix 1

Matrices

Figure A1.1 illustrates two coordinate systems in which the coordinates in the primed system are rotated through an angle α relative to the unprimed system. For an arbitrary point P we can easily find a relationship between the coordinates (x, y) in the original system and those (x', y') in the rotated system. From the geometry of the system we can write

$$x = x' \cos \alpha - y' \sin \alpha$$
$$y = x' \sin \alpha + y' \cos \alpha, \qquad (A1.1)$$

and we have a pair of linear equations which give the coordinate transformation between our two systems.

Sets of linear equations similar to those arise in applied mathematics under many different circumstances, and a special mathematical notation has been developed to handle such problems. We might write Eqs. (A1.1) in the form

$$r = \alpha r', \qquad (A1.2)$$

where r would be understood to be the vector to point P in the unprimed system, r' the vector to P in the primed system, and α some mathematical quantity that has the effect of transforming the vector r' into the vector r.

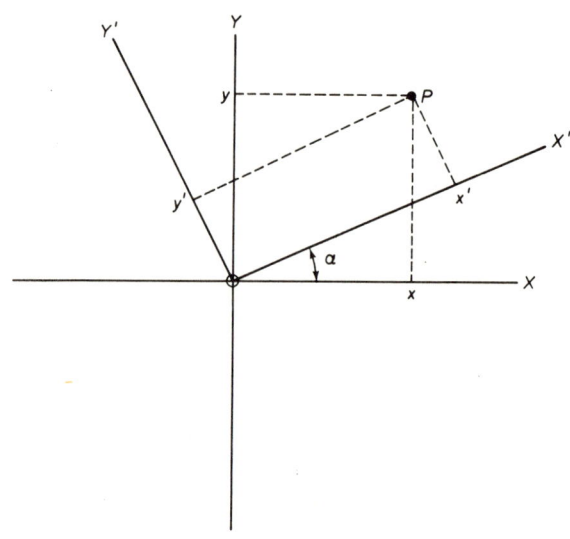

Fig. A1.1.

α is specified by the coefficients of x' and y' in Eqs. (A1.1). We now write α in the form

$$\alpha = \begin{bmatrix} \cos \alpha & -\sin \alpha \\ \sin \alpha & \cos \alpha \end{bmatrix}. \quad (A1.3)$$

Such an array of quantities (either constants or variables) is known as a matrix. The matrix is not a determinant. Matrices are generally shown in square brackets as above or in parentheses to distinguish them from determinants.

Matrices need not be square, that is, the number of rows and columns need not be equal. For example:

$$\begin{bmatrix} a & b & c \end{bmatrix}, \quad \begin{bmatrix} a \\ b \\ c \end{bmatrix}, \quad \begin{bmatrix} a & b & c \\ d & e & f \end{bmatrix}$$

are all matrices. The first two examples are given special names—row vector and column vector—the third example is simply a rectangular matrix.

We would now like to construct an algebra of matrices using an abstract matrix A defined by:

$$A = \begin{bmatrix} a_{11} & a_{12} & \cdots \\ a_{21} & a_{22} & \cdots \\ \vdots & \vdots & \cdots \end{bmatrix},$$

where A need not be square.

If we have a second matrix B defined by:

$$B = \begin{bmatrix} b_{11} & b_{12} & \cdots \\ b_{21} & b_{22} & \cdots \\ \vdots & \vdots & \cdots \end{bmatrix},$$

and if an equation such as:

$$A \cdot x + B \cdot x = (A+B) \cdot x$$

is satisfied, that is to say, if A and B satisfy the distributive law when operating on x, then we have defined matrix addition. For such an expression to have meaning it is necessary that both A and B have the same number of rows and columns and likewise their sum will have the same dimensions. As an example, consider the matrices A and B given by

$$A = \begin{bmatrix} 1 & 2 & 0 \\ 0 & 2 & 1 \end{bmatrix}, \quad B = \begin{bmatrix} 0 & 1 & 0 \\ -1 & 1 & 1 \end{bmatrix}. \tag{A1.4}$$

Their sum $C = A+B$ is given by adding the corresponding elements from A and B so that we have:

$$C = \begin{bmatrix} 1 & 3 & 0 \\ -1 & 3 & 2 \end{bmatrix}.$$

Matrix subtraction is simply performed in the same way so that

$$A - B = D = \begin{bmatrix} 1 & 1 & 0 \\ 1 & 1 & 0 \end{bmatrix}.$$

As in any subtraction, the ordering is important and $A - B = -(B - A)$.

Scalar multiplication of A by k is carried out by multiplying each element of A by k so that

$$3A = \begin{bmatrix} 3 & 6 & 0 \\ 0 & 6 & 3 \end{bmatrix}$$

using A as given in (A1.4)

If we now have a transformation generated by the following set of equations:

$$\begin{aligned} x_1 &= a_{11} y_1 + a_{12} y_2 \\ x_2 &= a_{21} y_1 + a_{22} y_2 \\ x_3 &= a_{31} y_1 + a_{32} y_2, \end{aligned} \tag{A1.5}$$

this could be represented in matrix notation as*

---

* It is common practice to use capital letter as a general notation for the matrix whose elements are given by the corresponding small letter. Occasionally A as above may be written as $[a_{ij}]$ or $[a_{ij}]_{mn}$, where m and n give the number of rows and columns, respectively.

$$x = Ay.$$

The elements y in the set of equations may, in fact, be related to a third set of elements given by

$$y = Bz$$

which in extended form is written as:

$$y_1 = b_{11}z_1 + b_{12}z_2$$
$$y_2 = b_{21}z_1 + b_{22}z_2. \tag{A1.6}$$

It would be convenient to be able to write $x = Ay$ as

$$x = ABz.$$

If we substitute Eqs. (A1.6) into Eqs. (A1.5), we get

$$x_1 = (a_{11}b_{11} + a_{12}b_{21})z_1 + (a_{11}b_{12} + a_{12}b_{22})z_2$$
$$x_2 = (a_{21}b_{11} + a_{22}b_{21})z_1 + (a_{21}b_{12} + a_{22}b_{22})z_2$$
$$x_3 = (a_{31}b_{11} + a_{32}b_{21})z_1 + (a_{31}b_{12} + a_{32}b_{22})z_2$$

which is clearly of the form $x = Cz$. We call C the product of A and B and write $C = AB$.

Any particular element c_{rt} or C can be seen to be of the form

$$c_{rt} = \sum_i a_{ri} b_{it}, \tag{A1.7}$$

and if the matrices A, B, and C are written out:

$$A = \begin{bmatrix} a_{11} & a_{12} \\ a_{21} & a_{22} \\ a_{31} & a_{32} \end{bmatrix}$$

and

$$B = \begin{bmatrix} b_{11} & b_{12} \\ b_{21} & b_{22} \end{bmatrix}$$

and finally

$$C = \begin{bmatrix} a_{11}b_{11} + a_{12}b_{21} & a_{11}b_{12} + a_{12}b_{22} \\ a_{21}b_{11} + a_{22}b_{21} & a_{21}b_{12} + a_{22}b_{22} \\ a_{31}b_{11} + a_{32}b_{21} & a_{31}b_{12} + a_{32}b_{22} \end{bmatrix}.$$

We can easily see that Eq. (A1.7) is merely a statement of the fact that matrix products are formed by the combination of the elements of a row of A with the elements of a column of B. It is important to understand clearly that the element in the ith row jth column of C is formed from the elements of the ith row of A and the jth column of B.

As an example we might consider the matrix product AB where

$$A = \begin{bmatrix} 1 & 2 \\ 2 & 1 \\ 1 & 3 \end{bmatrix}, \quad B = \begin{bmatrix} 1 & 1 \\ 2 & 2 \end{bmatrix}.$$

The product $C = AB$ is given by:

$$AB = C = \begin{bmatrix} 5 & 5 \\ 4 & 4 \\ 7 & 7 \end{bmatrix}.$$

The reader should verify this result before continuing with his reading.

A number of points concerning the process of matrix multiplication must now be made. First, the product can only be formed if the matrices are "conformable," i.e., if the number of columns of the first is equal to the number of rows of the second. If A is m by n and B is n by t, then the product C is m by t. Second, it follows from the definition of matrix multiplication that in general $AB \ne BA$, i.e., the matrices do not in general commute. In the example above, the product BA cannot be formed since in this order the matrices A and B do not conform.

As an example of the noncommutation of matrices, consider the following two square matrices:

$$A = \begin{bmatrix} 1 & 3 \\ 2 & 4 \end{bmatrix}, \quad B = \begin{bmatrix} 1 & 2 \\ 3 & 4 \end{bmatrix}.$$

Square matrices always conform in either order if they are of the same dimension. The products AB and BA are given by

$$AB = \begin{bmatrix} 9 & 14 \\ 14 & 20 \end{bmatrix}, \quad BA = \begin{bmatrix} 5 & 11 \\ 11 & 24 \end{bmatrix},$$

and clearly $AB \ne BA$ in spite of the fact that the matrices conform.

Extended matrix products, i.e., products of the form $ABC...F$ can be formed provided that they are in sequence conformable. In forming extended products, the ordering must remain fixed since the product is order dependent.

Special Matrices

1. *The Unit Matrix.* Unit multipliers have the effect of leaving the multiplicand unchanged. In matrix multiplication such a matrix takes

the form:

$$\begin{bmatrix} 1 & 0 & 0 & \cdots \\ 0 & 1 & 0 & \cdots \\ 0 & 0 & 1 & \cdots \\ & & \cdots & \end{bmatrix},$$

i.e., the unit multiplier is a square matrix with diagonal elements unity and all other elements zero. The unit matrix is usually denoted by I so that we can have

$$AI = A \text{ or } IA = A$$

since I does commute. However, if A is not square, the I's in these two cases will be of different orders so as to conform with A.

2. *Row and Column Vectors.* Matrices with only a single row of elements such as

$$[a \quad b \quad c \quad \cdots]$$

are termed row vectors, while those of the form

$$\begin{bmatrix} a \\ b \\ c \\ \vdots \end{bmatrix}$$

are called column vectors. These special matrices play an important role in many problems in applied mathematics and in the theory developed in this book.

A Useful Theorem

Any square matrix has associated with it a value known as its determinant. This determinant can be evaluated by the usual rules of linear algebra. In general we can state:

> *The determinant of the product of two square matrices A and B is equal to the product of the determinants of these matrices and similarly this rule holds for the product of any number of square matrices.*

We will simply verify this rule here for the 2×2 case. The reader is urged to examine the general proof as given in most texts on matrices and linear algebra. Consider the pair of matrices A and B:

$$A = \begin{bmatrix} a_{11} & a_{12} \\ a_{21} & a_{22} \end{bmatrix} \quad \text{and} \quad B = \begin{bmatrix} b_{11} & b_{12} \\ b_{21} & b_{22} \end{bmatrix}.$$

Their product is the matrix AB and is given by

$$AB = \begin{bmatrix} a_{11}b_{11}+a_{12}b_{21} & a_{11}b_{12}+a_{12}b_{22} \\ a_{21}b_{11}+a_{22}b_{21} & a_{21}b_{12}+a_{22}b_{22} \end{bmatrix}.$$

The determinants of A, B, and AB are

$$|A| = a_{11}a_{22} - a_{12}a_{21}$$
$$|B| = b_{11}b_{22} - b_{12}b_{21}$$
$$\begin{aligned} |AB| &= (a_{11}b_{11}+a_{12}b_{21})(a_{21}b_{12}+a_{22}b_{22}) - (a_{11}b_{12}+a_{12}b_{22})(a_{21}b_{11}+a_{22}b_{21}) \\ &= a_{11}a_{22}b_{11}b_{22} + a_{12}b_{21}a_{21}b_{12} - a_{11}b_{12}a_{22}b_{21} - a_{12}a_{21}b_{11}b_{22} \\ &= (a_{11}a_{22} - a_{12}a_{21})(b_{11}b_{22} - b_{12}b_{21}) \end{aligned}$$

which is by inspection $|A|\cdot|B|$. This can be extended to any number of 2×2 matrices by mathematical induction.

Appendix 2

Mathematical Theory of Aberrations

To develop some measure of the relative importance of the various aberrations in a given optical system, it is necessary to look at the fundamental form of the various Seidel sums and to fit them into the matrix theory. Consider Fig. A.2.1, where P_i is the image of a point source P, and P_i lies in the image plane at (x_p, y_p). We define a second reference plane by (x', y'). If an optical system is perfect, then the set of rays of fixed optical length coming from the source P generates a spherical surface centered about P_i, the size of which (in area) is determined by the appertures of the system. If our reference plane is the exit pupil (a particularly convenient choice of reference), then one spherical surface can, for example, be taken with a radius equal to the distance from $P_i(x_p, y_p)$ to the optic axis at the center of the exit pupil (along the principal ray).

Our ideal sphere will be centered about P_i and will have the form

$$(x_i' - x_P)^2 + (y_i' - y_P)^2 + z_i'^2 = r^2, \tag{A2.1}$$

while the real situation will be one in which the surface will deviate from the ideal spherical form above and we will choose to write the real surface in the form

$$(x_i' - x_P)^2 + (y_i' - y_P)^2 + z_i'^2 = (r + \Delta)^2, \tag{A2.2}$$

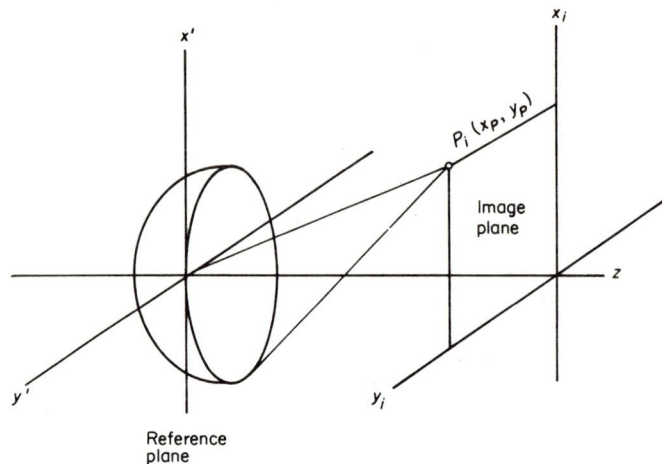

Fig. A2.1

where $\varDelta$ is some function that is small compared to r so that we can neglect terms in $\varDelta^2$. On expansion Eq. (A2.2) then takes the form

$$(x_i' - x_P)^2 + (y_i' - y_P)^2 + z_i'^2 = r^2 + 2r\varDelta. \tag{A2.3}$$

There are numerous ways in which we can express $\varDelta$. We can do this in a power series on a number of different coordinate systems, or in an expansion on the spherical harmonics, etc. Here we will treat the power series case. We will retain the assumption of rotational symmetry for our system and use polar coordinates in the reference planes defined by

$$\begin{aligned} x_P &= \rho_i \sin \alpha & x_1' &= r_1' \sin \vartheta \\ y_P &= \rho_i \cos \alpha & y_1' &= r_1' \cos \vartheta. \end{aligned} \tag{A2.4}$$

The assumption of rotational symmetry limits the functional form of $\varDelta$ to those terms that are invariant when the system is rotated, and by group theoretical arguments these are seen to be*

$$\varDelta[r_1'^2, \rho_i^2, \rho_i r_1' \cos(\vartheta - \alpha)].$$

We define the new coordinates ζ_1', ζ_2', ζ_3' by

---

* As an example, assume that $\varDelta$ depends on ρ_i. If the system is rotated through 180°, $\rho_i \to -\rho_i$ and $\varDelta$ would change in value of sign. Likewise a dependence only on $\cos \vartheta$ would produce a transformation in which $\cos \vartheta \to \cos(180° - \vartheta) = -\cos \vartheta$ for a 180° rotation of the system and thus produce a forbidden change in the sign or shape of $\varDelta$ with the rotation.

$$2\zeta_1' = \rho_i^2$$
$$\zeta_2' = r_1'\rho_i \cos(\varphi - \alpha) \tag{A2.5}$$
$$2\zeta_3' = r_1'^2$$

and expand Δ in a power series on these coordinates:

$$\Delta = a_0 + b_0\zeta_1' + b_1\zeta_2' + b_2\zeta_3' + \frac{1}{2}c_0\zeta_1'^2 + c_1\zeta_1'\zeta_2' + c_2\zeta_1'\zeta_3'$$
$$+ \frac{1}{2}c_3\zeta_2'^2 + c_4\zeta_2'\zeta_3' + \frac{1}{6}d_0\zeta_1'^3 + \cdots \tag{A2.6}$$

and the a_i, b_i, c_i, d_i, etc., are the quantities which characterize the optical system being treated. If we replace Δ in Eq. (A2.3) by the form (A2.6), we get the equation for the optical surface within our assumption that the deviation from sphericity is linear in Δ.

We now need the normals to this surface which are the rays at a given point (x_1', y_1') on the optical surface. By the usual rules of analytic geometry these are

$$\frac{x_1' - x_i}{2(x_1' - x_P) - 2r\frac{\delta\Delta}{\delta x_1'}} = \frac{y_1' - y_i}{2(y_1' - y_P) - 2r\frac{\delta\Delta}{\delta y_1'}} = \frac{z_1' - z_i}{2z_1'}, \tag{A2.7}$$

from which we can find

$$x_i\left(x_P, x_1', z_i, z_1', r, \frac{\delta\Delta}{\delta x_1'}\right)$$

and

$$y_i\left(y_P, y_1', z_i, z_1', r, \frac{\delta\Delta}{\delta y_1'}\right).$$

We consider the case where the image point is near the axis (x_P and y_P small) and where $\delta\Delta/\delta x_1'$ and $\delta\Delta/\delta y_1'$ are small (the real optical surface is smooth and changes slowly). We are especially interested in the region near the paraxial image point, thus we take points where z_i is small. Neglecting terms of the second order in small quantities, we find

$$x_i = x_P + z_i\frac{x_1'}{r} + r\frac{\delta\Delta}{\delta x_1'}$$
$$y_i = y_P + z_i\frac{y_1'}{r} + r\frac{\delta\Delta}{\delta y_1'} \tag{A2.8}$$

and in the paraxial image plane

$$x_i = x_P + r\frac{\delta\Delta}{\delta x_1'}$$
$$y_i = y_P + r\frac{\delta\Delta}{\delta y_1'},$$
(A2.9)

so that once we know $\delta\Delta/\delta x_1'$ and $\delta\Delta/\delta y_1'$ we know the image point in the paraxial image plane for the part of the optical surface in the region of (x_1', y_1'). From the definitions of (A2.4) we have

$$\frac{\delta\zeta_1}{\delta x_1'} = 0 \qquad \frac{\delta\zeta_1}{\delta y_1'} = 0$$
$$\frac{\delta\zeta_2}{\delta x_1'} = x_P \qquad \frac{\delta\zeta_2}{\delta y_1'} = y_P \qquad (A2.10)$$
$$\frac{\delta\zeta_3}{\delta x_1'} = x_1' \qquad \frac{\delta\zeta_3}{\delta y_1'} = y_1'.$$

and differentating (A2.6) we get

$$\frac{\delta\Delta}{\delta x_1'} = b_1 x_P + b_2 x_1' + c_1 \zeta_1 x_P + c_2 \zeta_1 x_1' + c_3 \zeta_2 x_1' + c_4 \zeta_2 x_1'$$
$$+ c_4 \zeta_3 x_P + c_5 \zeta_3 x_1'$$
$$\frac{\delta\Delta}{\delta y_1'} = b_1 y_P + b_2 y_1' + c_1 \zeta_1 y_P + c_2 \zeta_1 y_1' + c_3 \zeta_2 y_1' + c_4 \zeta_2 y_1'$$
$$+ c_4 \zeta_3 y_P + c_5 \zeta_3 y_1'.$$
(A2.11)

These can now be introduced into (A2.9) and we have the intercept in the paraxial image plane for the ray arising on the optical surface in the region of (x_1', y_1'). The constants b_1 and b_2 may be evaluated by comparing the expanded Eqs. (A2.9) with the ideal case, and we find $b_1 = 0$ and $b_2 = z/r^2$. The constants c_i in Eqs. (A2.11) now fix the image defects of the system.

We want to examine aberrations in terms of the object rather than in terms of the image since we are primarily interested in the effect on the image of changes in the object position so that in reality we should write the aberration equations in terms of the object point. The object point will have coordinates (x_0, y_0), and coordinates in the entrance conjugate to (x', y') in the exit pupil reference will be given by (x, y). The transformation between (x_i, y_i) and (x_0, y_0), (x, y), and (x', y') are given in our paraxial theory by

$$x_i = \mu x_0 \qquad y_i = \mu y_0$$
$$x_1' = \mu_1 x_1 \qquad y_1' = \mu_1 y_1,$$
(A2.12)

where μ_1 is the magnification between the pupils. In the image plane we now find

$$x_i = \mu x + S_{10}\zeta_1 x_0 + S_{20}\zeta_1 x_1 + S_{11}\zeta_2 x_0 + S_{21}(\zeta_2 x + \zeta_3 x_0) + S_{22}\zeta_3 x$$
$$y_i = \mu y + S_{10}\zeta_1 y_0 + S_{20}\zeta_2 y + S_{11}\zeta_2 y_0 + S_{21}(\zeta_2 y + \zeta_3 y_0) + S_{22}\zeta_3 y,$$
(A2.13)

where we have grouped constant terms, and

$$S_{10} = rc_1 \mu^3$$
$$S_{20} = rc_2 \mu^2 \mu_1$$
$$S_{11} = rc_3 \mu^2 \mu_1$$
$$S_{21} = rc_4 \mu \mu_1^2$$
$$S_{22} = rc_5 \mu_1^3,$$

and ζ_i is conjugate to ζ_i', etc.

The various aberrations arise in the following way:

Spherical aberration	$S_{22} \neq 0$
Coma	$S_{21} \neq 0$
Astigmatism	$S_{11} \neq 0$
Field curvature	$S_{20} \neq 0$
Distortion	$S_{10} \neq 0$.

In order to find the S's, we return to the paraxial solution (3.11):

$$\begin{pmatrix} \lambda' \\ h' \end{pmatrix} = \begin{pmatrix} \dfrac{1}{\mu} & -a \\ 0 & \mu \end{pmatrix} \begin{pmatrix} \lambda \\ h \end{pmatrix}.$$
(A2.14)

The series expansions of λ' and h' (third-order aberrations) will have the form

$$\lambda' = \frac{1}{\mu}\lambda - ah + d_1\lambda^3 + d_2\lambda^2 h + d_3\lambda h^2 + d_4 h^3$$
(A2.15a)

$$h' = \mu h + e_1\lambda^3 + e_2\lambda^2 h + e_3\lambda h^2 + e_4 h^3,$$
(A2.15b)

which can be written in matrix form as

$$\begin{pmatrix} \lambda' \\ h' \\ \lambda'^3 \\ \lambda'^2 h' \\ \lambda' h'^2 \\ h'^3 \end{pmatrix} = \begin{pmatrix} \dfrac{1}{\mu} & -a & d_1 & d_2 & d_3 & d_4 \\ 0 & \mu & e_1 & e_2 & e_3 & e_4 \\ 0 & 0 & \dfrac{1}{\mu^3} & -3\dfrac{a}{\mu^2} & 3\dfrac{a^2}{\mu} & -a^3 \\ 0 & 0 & 0 & \dfrac{1}{\mu} & 2a & a^2\mu \\ 0 & 0 & 0 & 0 & \mu & -a\mu^2 \\ 0 & 0 & 0 & 0 & 0 & \mu^3 \end{pmatrix} \begin{pmatrix} \lambda \\ h \\ \lambda^3 \\ \lambda^2 h \\ \lambda h^2 \\ h^3 \end{pmatrix}, \quad \text{(A2.16)}$$

where the additional terms below the second row are found by forming appropriate product $\lambda'^2 h'$; for example, by squaring Eq. (A2.15a) and then multiplying by Eq. (A2.15b), rejecting terms of order greater than three.

Perhaps a representation more useful than (A2.16) can be developed using (A2.12) with the appropriate variables used there. We get matrices

$$\begin{pmatrix} x_i \\ x' \\ \zeta_1' x_i \\ \zeta_1' x_1' \\ \zeta_2' x_i \\ \zeta_2' x_1' \\ \zeta_3' x_i \\ \zeta_3' x_1' \end{pmatrix} = \begin{pmatrix} \mu & 0 & S_{10} & S_{20} & S_{11} & S_{21} & S_{21} & S_{22} \\ 0 & \mu_1 & D_{10} & D_{20} & D_{11} & D_{21} & D_{21} & D_{22} \\ 0 & 0 & \mu^3 & 0 & 0 & 0 & 0 & 0 \\ 0 & 0 & 0 & \mu_1 \mu^2 & 0 & 0 & 0 & 0 \\ 0 & 0 & 0 & 0 & \mu_1 \mu^2 & 0 & 0 & 0 \\ 0 & 0 & 0 & 0 & 0 & \mu_1^2 \mu & 0 & 0 \\ 0 & 0 & 0 & 0 & 0 & 0 & \mu_1^2 \mu & 0 \\ 0 & 0 & 0 & 0 & 0 & 0 & 0 & \mu_1^3 \end{pmatrix} \begin{pmatrix} x_0 \\ x \\ \zeta_1 x_0 \\ \zeta_0 x \\ \zeta_2 x_0 \\ \zeta_2 x \\ \zeta_3 x_0 \\ \zeta_3 x \end{pmatrix},$$

and for the y's:

$$\begin{pmatrix} y_i \\ y' \\ \zeta_1' y_i \\ \zeta_1' y_1' \\ \zeta_2' y_i \\ \zeta_2' y_1' \\ \zeta_3' y_i \\ \zeta_3' y_1' \end{pmatrix} = \begin{pmatrix} \mu & 0 & S_{10} & S_{20} & S_{11} & S_{21} & S_{21} & S_{22} \\ 0 & \mu_1 & D_{10} & D_{20} & D_{11} & D_{21} & D_{21} & D_{22} \\ 0 & 0 & \mu^3 & 0 & 0 & 0 & 0 & 0 \\ 0 & 0 & 0 & \mu_1 \mu^2 & 0 & 0 & 0 & 0 \\ 0 & 0 & 0 & 0 & \mu_1 \mu^2 & 0 & 0 & 0 \\ 0 & 0 & 0 & 0 & 0 & \mu_1^2 \mu & 0 & 0 \\ 0 & 0 & 0 & 0 & 0 & 0 & \mu_1^2 \mu & 0 \\ 0 & 0 & 0 & 0 & 0 & 0 & 0 & \mu_1^3 \end{pmatrix} \begin{pmatrix} y_0 \\ y \\ \zeta_1 x_0 \\ \zeta_0 x \\ \zeta_2 y_0 \\ \zeta_2 y \\ \zeta_3 y_0 \\ \zeta_3 y \end{pmatrix}.$$

This illustrates two ways in which aberrations may be treated in the matrix theory. The actual calculation of the coefficients is beyond the scope of this book.

As an illustration, we recall that coma occurs when $S_{21} \neq 0$. The dependence of x_i on S_{21} is given by

$$x_i = S_{21}(\zeta_2 x + \zeta_3 x_0).$$

Using the definitions of ζ_i in Eqs. (A2.5) transformed onto the appropriate plane, we find

$$x_i = S_{21}(r\rho_0 x_0 \cos(\vartheta-\alpha)+\zeta_3 x_0),$$

and for points on the optic axis $\rho_0 = x_0 = 0$. This term vanishes regardless of the value of S_{21}, and we thereby verify the statement made in Chapter 7 that coma does not occur for points on or very near the optic axis.

Bibliography

Books of General Interest

Ditchburn, R. W., *Light*, 2nd ed., Wiley, New York, 1963.
Jenkins, F. A., and White, H. E., *Fundamentals of Optics*, 3rd ed., McGraw-Hill, New York, 1957.
Martin, L. C., *Geometrical Optics*, Pitman, London, 1955.
Martin, L. C., *Technical Optics*, Pitman, London, 1952.
Valasek, J., *Introduction to Theoretical and Experimental Optics*, Wiley, New York, 1949.
Welford, W. T., *Geometrical Optics*, North-Holland, Amsterdam, 1962.

Books Containing Treatments of the Matrix Theory

Brouwer, W., *Matrix Methods in Optical Instrumental Design*, Benjamin, New York, 1964.
Nussbaum, A., *Geometrical Optics, An Introduction*, Addison-Wesley, Reading, Mass., 1968.

Leatham, G. G., *The Elementary Theory of the Symmetrical Optical Instrument*, Hafner, New York, 1960.

O'Neill, E. L., *Introduction to Statistical Optics*, Addison-Wesley, Reading, Mass., 1963, Chapter 3.

Van Heel, A. C. S., (ed.), *Advanced Optical Techniques*, North-Holland, Amsterdam, 1967, Chapter 16.

Index

A

Aberrations, 77, 115
    astigmatism, 84, 119
    chromatic, 70
    coma, 83
    curvature of field, 87, 119
    distortion, 87, 119
    fifth-order, 88
    geometric, 77
Angle of deviation, 14
Angle of incidence, 5
Angle of reflection, 5
Angle of refraction, 6
Angular magnification, 96
Aperture stop, 64
Astigmatism, 84

B

Barrel distortion, 87
Binoculars, 97

C

Cameras, 101
Cemented doublet, 72
Changing the external medium, 49
Chromatic aberrations, 70
Circle of least confusion, 81
Column vector, 112
Coma, 83, 119
Condenser, 101
Conjugate points, 38
Convex mirror, 57
Cooke triplet, 101
Curvature of the field, 87, 119

D

Determinants, 29
    of system matrix, 30
Direction cosines, 17
Dispersion, 71
Distortion, 87, 119
Doublet lens, 72

E

Entrance pupil, 65
Erecting lens, 99
Exit pupil, 65
Eyepiece lens, 94
Eyepieces, 92

F

f-Number, 102
Fermat's principle, 2
Field lens, 94
Field of view, 69
Field stop, 64
Fifth-order aberrations, 88
Focal length, 11
Focal paraboloids, 86
Focal planes, 39
Focal points, 39
Fraunhofer lines, 71

G

Gallilean telescope, 96
Gaussian approximation, 1, 7, 21, 77
Gaussian constants, 37
Graphical image construction, 45

H

Huygen's eyepiece, 93

I

Image aberration, 9
Imaging by refraction, 11, 30
Index of refraction, 6

K

Kellner eyepiece, 93

L

Lateral chromatic aberration, 71
Lens equation, 11, 39
Lenses, in contact, 49
    with reflecting surface, 58
Lensmakers' equation, 11, 48
Longitudinal chromatic aberration, 71
Longitudinal spherical aberration, 80

M

Magnification, 32, 96
    angular, 96
    visual, 90
Matrix operations, 107
Microscope, 99
Mirrors, 53
Monoculars, 97

N

Nodal points, 43
Numerical aperture, 100

O

Objective (microscope), 100
Oil immersion objective, 100
Optical direction cosine, 19
Optical path, 3
Optical systems, 27
Optic axis, 7

P

Paraxial ray; see Gaussian approximation
Petzval surface, 87
Pin-cushion distortion, 87
Plane mirror, 13, 56
Power, 90
    of surface, 19
Principal planes, 37
Principal ray, 69
Projection systems, 105
Pupils, 63

R

Ramsden eyepiece, 73, 90, 93
Ray model, 1
Real image, 9
Reduced optical length, 20
Reflecting telescopes, 98
Reflection, 4
Refraction, 6
    at spherical surface, 7
    matrix, 16
Refractive index, 3, 6
    negative, 55
Rotational invariants, 116
Row vector, 112

S

Sagittal rays, 84
Seidel sums, 78, 115
Shadowing, 1
Shape factor, 81, 83
Simple magnifier, 89
Sign convention, 15
Snell's law, 6, 25
Spherical aberration, 80
Stops, 63
System matrix, 29

T

Tangential rays, 84
Telephoto lens, 103
Telescopes, 95
Tessar lens, 102
Thin lens, 10, 47
Translation matrix, 19
Triplet magnifier, 90

U

Unit matrix, 23, 111
    planes, 37
    points, 37

V

Vertex, 7
Vignetting, 69
Virtual image, 9
Visual magnification, 90

W

Weighted direction cosine, 19